CLEP

College Level Examination Program

**Science
Series**

Copyright © 2016
All rights reserved. No part of the material protected by this copyright notice may be reproduced or utilized in any form or by any means, electronic or mechanical, including photocopying or recording or by any information storage and retrievable system, without written permission from the copyright holder.

To obtain permission(s) to use the material from this work for any purpose including workshops or seminars, please submit a written request to:

XAMonline, Inc.
21 Orient Avenue
Melrose, MA 02176
Toll Free: 1-800-301-4647
Email: info@xamonline.com
Web: www.xamonline.com
Fax: 1-617-583-5552

Library of Congress Cataloging-in-Publication Data
Wynne, Sharon

CLEP Science Series/ Sharon Wynne
   ISBN: 978-1-60787-580-2

1. CLEP   2. Study Guides   3. Science

Disclaimer:
The opinions expressed in this publication are the sole works of XAMonline and were created independently from the College Board, or other testing affiliates. Between the time of publication and printing, specific test standards as well as testing formats and website information may change that are not included in part or in whole within this product. XAMonline develops sample test questions, and they reflect similar content as on real tests; however, they are not former tests. XAMonline assembles content that aligns with test standards but makes no claims nor guarantees candidates a passing score.

Cover photos provided by © Can Stock Photo Inc./maxxyustas/14557820; © Can Stock Photo Inc./kasta/28180375; © Can Stock Photo Inc./piai/6503471; © iStock.com/luchschen/46948150

Printed in the United States of America
CLEP Science Series
ISBN: 978-1-60787-580-2

TABLE OF CONTENTS

Natural Sciences 1

Sample Test 3

Answer Key 27

Rationales 28

Biology 87

Sample Test 90

Answer Key 114

Rationales 115

Chemistry 174

Sample Test 177

Answer Key 198

Rationales 199

NATURAL SCIENCES

Description of the Examination

The Natural Sciences examination covers a wide range of topics frequently taught in introductory courses surveying both biological and physical sciences at the freshman or sophomore level. Such courses generally satisfy distribution or general education requirements in science that usually are not required of nor taken by science majors. The Natural Sciences exam is not intended for those specializing in science; it is intended to test the understanding of scientific concepts that an adult with a liberal arts education should have. It does not stress the retention of factual details; rather, it emphasizes the knowledge and application of the basic principles and concepts of science, the comprehension of scientific information, and the understanding of issues of science in contemporary society.

The primary objective of the examination is to give candidates the opportunity to demonstrate a level of knowledge and understanding expected of college students meeting a distribution or general education requirement in the natural sciences. An institution may grant up to six semester hours (or the equivalent) of credit toward fulfillment of such a requirement for satisfactory scores on the examination. Some may grant specific course credit, on the basis of the total score for a two-semester survey course covering both biological and physical sciences.

The examination contains approximately 120 questions to be answered in 90 minutes. Some of these are pretest questions that will not be scored. Any time candidates spend on tutorials and providing personal information is in addition to the actual testing time.

Knowledge and Skills Required

The Natural Sciences examination requires candidates to demonstrate one or more of the following abilities in the approximate proportions indicated.

- Knowledge of fundamental facts, concepts, and principles (about 40 percent of the examination)
- Interpretation and comprehension of information (about 20 percent of the examination) presented in the form of graphs, diagrams, tables, equations, or verbal passages
- Qualitative and quantitative application of scientific principles (about 40 percent of the examination), including applications based on material presented in the form of graphs, diagrams, tables, equations, or verbal passages; more emphasis is given to qualitative than quantitative applications

Topical Specifications:

50%	**Biological Science**
10%	**Origin and evolution of life, classification of organisms**
10%	**Cell organization, cell division, chemical nature of the gene, bioenergetics, biosynthesis**
20%	**Structure, function, and development in organisms; patterns of heredity**
10%	**Concepts of population biology with emphasis on ecology**
50%	**Physical Science**
7%	**Atomic and nuclear structure and properties, elementary particles, nuclear reactions**
10%	**Chemical elements, compounds and reactions, molecular structure and bonding**
12%	**Heat, thermodynamics, and states of matter; classical mechanics; relativity**

NATURAL SCIENCES

4% **Electricity and magnetism, waves, light, and sound**
7% **The universe: galaxies, stars, the solar system**
10% **The Earth: atmosphere, hydrosphere, structure features, geologic processes, and history**

The examination includes some questions that are interdisciplinary and cannot be classified in one of the listed categories. Some of the questions cover topics that overlap with those listed previously, drawing on areas such as history and philosophy of science, scientific methods, science applications and technology, and the relationship of science to contemporary problems of society, such as environmental pollution and depletion of natural resources. Some questions are laboratory oriented.

NATURAL SCIENCES

SAMPLE TEST

DIRECTIONS: Read each item and select the best response

1. **According to scientists, what is the estimate age of the Earth?**

   A. 4.5 million years

   B. 4.5 billion years

   C. 450 million years

   D. 1.000 million years

   E. 10.000 million years

2. **The first cells that evolved on earth were probably of which type?**

   A. Autotrophic

   B. Eukaryotic

   C. Similar to viruses

   D. Prokaryotic

   E. Endosymbiotic

3. **What is a major principle of the Endosymbiotic Theory?**

   A. Birds and dinosaurs share a common ancestor.

   B. Animals evolved in close relationships with one another.

   C. Prokaryotes arose from eukaryotes.

   D. Inorganic compounds are the basis of living things.

   E. Eukaryotes arose from very simple prokaryotes.

4. **According to Oparin & Haldane's theory, the primitive atmosphere was composed by**

   A. Hydrogen, methane, water, ammonia

   B. Oxygen, methane, water, ammonia

   C. Oxygen and carbonic gas

   D. Oxygen, carbonic gas, nitrogen

   E. Hydrogen, methane, water, ozone

5. **Which of these is true about natural selection?**

   A. It acts on an individual genotype

   B. It is not currently happening

   C. It is only an animal phenomenon

   D. It acts on the individual phenotype

   E. It is used to prevent overpopulation

6. **Which of these is a result of reproductive isolation?**

   A. Extinction

   B. Migration

   C. Fossilization

   D. Speciation

   E. Radiation

NATURAL SCIENCES

7. Which of these is NOT a prezygotic barrier?

   A. Geographical isolation

   B. Hybrid sterility

   C. Temporal isolation

   D. Mechanical isolation

   E. Behavioral isolation

8. Which mode of natural selection favors the more common phenotypes?

   A. Directional selection

   B. Positive selection

   C. Stabilizing selection

   D. Diversifying selection

   E. Disruptive selection

9. Which phylum accounts for 85% of all animal species?

   A. Nematoda

   B. Chordata

   C. Arthropoda

   D. Cnidaria

   E. Annelida

10. The scientific name of humans is *Homo sapiens*. Choose the proper classification beginning with kingdom and ending with order

    A. Animalia, Vertebrata, Mammalia, Primates, Hominidae

    B. Animalia, Vertebrata, Chordata, Mammalia, Primates

    C. Animalia, Chordata, Vertebrata, Mammalia, Primates

    D. Chordata, Vertebrata, Primate, *Homo, sapiens*

    E. Chordata, Primates, Hominidae, *Homo, sapiens*

11. Which of the following animals is coelomate?

    I. Flatworms
    II. Earthworms
    III. Crickets

    A. I only

    B. II only

    C. III only

    D. I and III

    E. II and III

NATURAL SCIENCES

12. **Heterotrophic organisms that have cell walls with chitin are classified as**

    A. Plants

    B. Bacteria

    C. Fungi

    D. Animals

    E. Protists

13. **Of what are viruses made?**

    A. A protein coat surrounding a nucleic acid

    B. RNA and protein surrounded by a cell wall

    C. A nucleic acid surrounding a protein coat

    D. Protein surrounded by DNA

    E. A lipid bilayer surrounding a protein coat and RNA

14. **According to the fluid-mosaic model of the cell membrane, membranes are composed of**

    A. A phospholipid bilayer with proteins embedded in the layers

    B. One layer of phospholipids with cholesterol embedded in the layer

    C. Two layers of protein with lipids embedded in the layers

    D. DNA and fluid proteins

    E. Two layers of phospholipids and DNA

15. **Which of the following is not part of the cytoskeleton?**

    A. Vacuoles

    B. Microfilaments

    C. Microtubules

    D. Intermediate filaments

    E. Motor proteins

16. **Bacteria commonly reproduce by a process called binary fission. Which of the following best defines this process?**

    A. Viral vectors carry DNA to new bacteria

    B. DNA from one bacterium enters another

    C. DNA doubles and the bacterial cell divides

    D. DNA from dead cells is absorbed into bacteria

    E. Bacteria merge with others to form new species

17. What is the stage of mitosis shown in the diagram?

    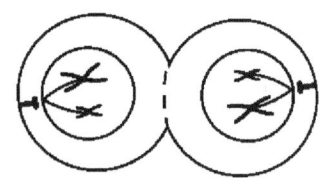

    A. Prophase

    B. Telophase

    C. Anaphase

    D. Metaphase

    E. Interphase

18. Crossing over, which increases genetic diversity, occurs during which stage of meiosis?

    A. Telophase II in meiosis

    B. Metaphase in mitosis

    C. Interphase in meiosis

    D. Prophase I in meiosis

    E. Metaphase II in meiosis

19. What is the enzyme that unwinds DNA during replication?

    A. DNAse

    B. Replicase

    C. DNA helicase

    D. DNA topoisomerases

    E. DNA polymerase

20. Which part of a DNA nucleotide can vary?

    A. Deoxyribose

    B. Phosphate group

    C. Hydrogen bonds

    D. Sugar

    E. Nitrogenous base

21. Which of these describes facilitated diffusion?

    A. It requires energy

    B. It only happens in plant cells

    C. It only allows molecules to leave a cell but not to enter it

    D. It produces a significant amount of energy for the cell

    E. It needs a transport molecule to pass through the membrane

22. During which part of photosynthesis is oxygen given off?

    A. Light reactions

    B. Dark reactions

    C. Krebs cycle

    D. Reduction of NAD+ to NADH

    E. Phosphorylation

23. During the Krebs cycle, 8 carrier molecules are formed. What are they?

   A. 3 NADH, 3 FADH, 2 ATP

   B. 6 NADH and 2 ATP

   C. 4 $FADH_2$ and 4 ATP

   D. 6 NADH and 2 $FADH_2$

   E. 4 NADH and 4 $FADH_2$

24. In the comparison of respiration to photosynthesis, which statement is true?

   A. Oxygen is a waste product in photosynthesis but not in respiration

   B. Glucose is produced in respiration but not in photosynthesis

   C. Carbon dioxide is formed in photosynthesis but not in respiration

   D. Water is formed in respiration but not in photosynthesis

   E. Carbon dioxide and water are formed in photosynthesis

25. Identify the correct sequence of organization of living things

   A. Cell – organelle – organ – tissue – organ system – organism

   B. Cell – tissue – organ – organelle – organ system – organism

   C. Organelle – cell – tissue – organ – organ system – organism

   D. Organ system – tissue – organelle – cell – organism – organ

   E. Organism – organ system – tissue – cell – organelle – organ

26. Bird wings, human arms, and the flipper of whales have the same bone structure and different functions. These are called

   A. Polymorphic structures

   B. Homologous structures

   C. Vestigial structures

   D. Analogous structures

   E. Primitive structure

27. Parts of the nervous system include all but the following

   A. Brain

   B. Spinal cord

   C. Axons

   D. Venules

   E. Glial cells

NATURAL SCIENCES

28. Which of the following best completes the statement below? Peristalsis and the movement of the iris are possible due to the action of _____ muscles.

    A. Skeletal

    B. Smooth

    C. Cardiac

    D. Striated

    E. Voluntary

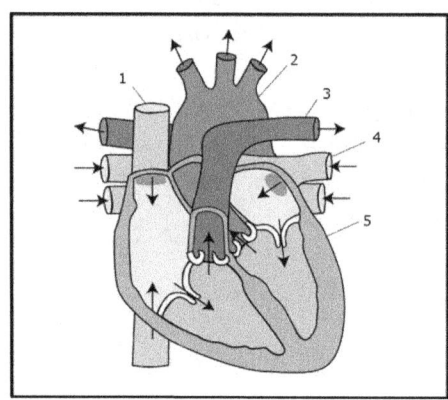

29. Consider the heart illustration above, with arrows indicating the direction of blood flow. Which number indicates the aorta?

    A. 1

    B. 2

    C. 3

    D. 4

    E. 5

30. Microorganisms use all but which of the following for locomotion?

    A. Pseudopods

    B. Flagella

    C. Cilia

    D. Pili

    E. Villi

31. Which of these is a function of the cardiovascular system?

    A. Move oxygenated blood around the body

    B. Oxygenate the blood through gas exchange

    C. Act as an exocrine system

    D. Flush toxins out of the body

    E. Transport signals from the brain

32. Which of these steroids is not created in the gonads?

    A. Testosterone

    B. Estrogen

    C. Progesterone

    D. ACTH

    E. FSH

NATURAL SCIENCES

33. **The role of neurotransmitters in nerve action is**

    A. To turn off the sodium pump

    B. To turn off the calcium pump

    C. To send impulses to neurons

    D. To send impulses to the body

    E. To maintain the membrane polarized

34. **Which of the following is NOT a function of the vertebrate skin**

    A. Respiration

    B. Protection

    C. Sensation

    D. Regulation of temperature

    E. Regulation of reproduction

35. **Homeostatic mechanisms in the body do NOT include**

    A. Thermoregulation

    B. Excretion

    C. Respiration

    D. Osmoregulation

    E. Hemostasis

36. **What controls gas exchange on the bottom of a plant leaf?**

    A. Stomata

    B. Epidermis

    C. Collenchyma and schlerenchyma

    D. Palisade mesophyll

    E. Trichomes

37. **Hormones are essential to the regulation of reproduction. What organ is responsible for the release of hormones for sexual maturity?**

    A. Thymus gland

    B. Hypothalamus

    C. Pancreas

    D. Thyroid gland

    E. Cerebellum

38. **Fertilization in humans usually occurs in the**

    A. Uterus

    B. Ovary

    C. Fallopian tubes

    D. Vagina

    E. Cervix

39. After sea turtles hatch on the beach, they start the journey to the ocean. This is due to

A. Learned behavior

B. Territoriality

C. The tide

D. Innate behavior

E. Feeding strategy

40. What is any foreign particle called that causes an immune reaction?

A. An antigen

B. A histocompatibity complex

C. An antibody

D. A vaccine

E. A bacteriophage

41. What is (are) the germ layer(s) missing in diploblastic animals?

A. Ectoderm only

B. Mesoderm only

C. Endoderm only

D. Ectoderm and mesoderm

E. Endoderm and mesoderm

42. What is the order of the stages that happen after fertilization?

A. Blastula – gastrulation – neurulation – organogenesis – cleavage

B. Cleavage – neurulation – gastrulation – organogenesis – blastula

C. Cleavage – blastula – gastrulation – neurulation – cell growth

D. Cell growth – gastrulation – blastula – neurulation – organogenesis

E. Cleavage – blastula – gastrulation – neurulation – organogenesis

43. What is the general term for a change that affects the sequence of bases in a gene?

A. Deletion

B. Polyploidy

C. Mutation

D. Duplication

E. Substitution

NATURAL SCIENCES

44. **What can be said about homozygous individuals?**

    A. They have two different alleles

    B. They are of the same species

    C. They exhibit the same features

    D. They have a pair of identical alleles

    E. They produce identical offspring

45. **In a Punnett square with a single trait, what are the ratios of genotypes produced between two heterozygous individuals?**

    A. 1:2:2

    B. 2:1:1

    C. 1:1:1

    D. 1:2:1

    E. 2:2:2

46. **A child with type O blood has a father with type A blood and a mother with type B blood. The genotypes of the parents respectively would be which of the following?**

    A. AA and BO

    B. AO and BO

    C. AA and BB

    D. AO and OO

    E. OO and BO

47. **Which of these defines the Law of Segregation defined by Gregor Mendel?**

    A. After meiosis, each new cell will contain an allele that is recessive.

    B. Only one of two alleles is expressed in a heterozygous organism.

    C. The allele expressed is always the dominant allele.

    D. Alleles of one trait do not affect the inheritance of alleles on another chromosome.

    E. When sex cells form, the two alleles that determine a trait will end up on different gametes.

48. **Hemophilia and color-blindness are examples of**

    A. Lethal alleles

    B. Codominance system

    C. Sex-linked traits

    D. Incomplete dominance

    E. Nondisjunction

NATURAL SCIENCES

49. Which of the following is NOT an abiotic factor?

    A. Temperature

    B. Rainfall

    C. Soil quality

    D. Predation

    E. Wind speed

50. An experiment was performed to measure the growth of bacteria at different temperatures. The cultures were kept on a 12 hour light/dark cycle and given the same amount of nutrients. Which of these is the independent variable?

    A. Growth of number of colonies

    B. Amount of nutrients

    C. Type of bacteria used

    D. Light duration

    E. Temperature

51. Which term is not associated with the water cycle?

    A. Precipitation

    B. Transpiration

    C. Fixation

    D. Evaporation

    E. Infiltration

52. Which trophic level has the highest ecological efficiency?

    A. Decomposers

    B. Producers

    C. Tertiary consumers

    D. Secondary consumers

    E. Primary consumers

53. What is NOT true about competition?

    A. May occur between very different species

    B. It is usually asymmetric, affecting one species more than the other

    C. It increases the amount of available resources

    D. May affect the abundance of competitors

    E. Competition is a common process in natural communities

54. Which of the following is true about parasites?

   I. All parasites are facultative
   II. Parasites can be either ecto- or endoparasites
   III. Parasites increase their hosts' fitness
   IV. Parasites always kill their hosts

   A. I only
   B. II only
   C. I and II
   D. I, II and III
   E. I, II, III and IV

55. A clownfish is protected by a sea anemone's tentacles, and in turn, the anemone receives uneaten food from the clownfish. What type of symbiosis is exemplified by this example?

   A. Mutualism
   B. Parasitism
   C. Commensalism
   D. Competition
   E. Amensalism

56. Since the industrial revolution, the size of the human population has been

   A. Decreasing
   B. Stable
   C. Increasing slowly
   D. Changing randomly
   E. Increasing exponentially

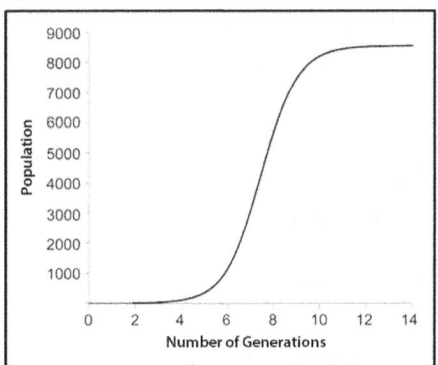

57. According to the graphic above, what is the approximate population size after 8 generations?

   A. 1000
   B. 2000
   C. 6000
   D. 8000
   E. 12000

NATURAL SCIENCES

58. Xerophytes are more commonly found in which biome?

    A. Temperate forests

    B. Tropical forests

    C. Oceans

    D. Deserts

    E. Grasslands

59. Which zone of the ocean receives no light and has the lowest temperatures?

    A. Intertidal

    B. Pelagic

    C. Abyssal

    D. Euphotic

    E. Epipelagic

60. Which biome is the most prevalent on Earth?

    A. Marine

    B. Desert

    C. Savanna

    D. Tundra

    E. Lakes and rivers

61. How many neutrons are there in $^{60}_{27}Co$?

    A. 27

    B. 33

    C. 60

    D. 87

    E. 14

62. Which of the following can be determined from the periodic table?

    I. The number of protons
    II. The number of neutrons
    III. The number of isotopes of that atom
    IV. The number of valence electrons

    A. I only

    B. I and II

    C. I, II, and III

    D. I, II, III, IV

    E. I, II and IV

63. When a radioactive material emits an alpha particle only, its atomic number

    A. Decreases

    B. Increases

    C. Remains unchanged

    D. Changes randomly

    E. Approaches zero

NATURAL SCIENCES

64. The terrestrial composition of an element is 50.7% as a stable isotope with an atomic mass of 78.9 u and 49.3% as a stable isotope with an atomic mass of 80.9 u. What is the atomic mass of the element?

A. 79.0 u

B. 79.8 u

C. 79.9 u

D. 80.8 u

E. 80.0 u

65. Moving down a column on the Periodic Table

I. The atomic radius increase
II. Ionization energy increase
III. Protons are added
IV. Metallic characteristics increase

A. I only

B. III only

C. I and II

D. III and IV

E. I, II, III

66. Which of the following quantum numbers are needed to define the position of the electrons in an element?

A. Principal, circular, magnetic and electromagnetic

B. Principal, angular momentum, magnetic, and spin

C. Angular, magnetic, electronic, spin

D. Primary, angular momentum, magnetic, and spin

E. Principal only

67. $^{3}_{1}H$ decays with a half-life of 12 years. Twenty four years ago, 3.0 g of pure $^{3}_{1}H$ were placed in sealed container. How many grams of $^{3}_{1}H$ remain?

A. 0.38 g

B. 0.75 g

C. 1.5 g

D. 0.125 g

E. 3.0 g

NATURAL SCIENCES

68. Write a balanced nuclear equation for the emission of an alpha particle by polonium-209

    A. $^{209}_{84}Po \rightarrow\ ^{205}_{81}Po +\ ^{4}_{2}He$

    B. $^{209}_{84}Po \rightarrow\ ^{205}_{82}Po +\ ^{4}_{2}He$

    C. $^{209}_{84}Po \rightarrow\ ^{209}_{85}Po +\ ^{0}_{-1}He$

    D. $^{209}_{84}Po \rightarrow\ ^{205}_{82}Po +\ ^{4}_{2}He$

    E. $^{209}_{84}Po \rightarrow\ ^{209}_{83}Po +\ ^{4}_{2}He$

69. Based on trends in the periodic table, which of the following properties would you expect to be greater for Rb than for K?

    I. Density
    II. Melting point
    III. Ionization energy
    IV. Oxidation number in a compound with chlorine

    A. I only

    B. I, II, and III

    C. II and III

    D. I and IV

    E. I, II, III, and IV

70. According to the periodic table, which list contains only metals?

    A. Li, Cd, Ca, S

    B. Li, Cd, Ca, He

    C. Li, Cd, Ca, Pb

    D. Li, C, Ca, He

    E. Ca, Fe, F, Pb

71. Which of the following are the properties of noble gases?

    I. They are colorless and odorless under STP
    II. They have little tendency to gain or lose electrons
    III. High melting points
    IV. Full valence electron shells

    A. I and II

    B. I and III

    C. II and III

    D. I, II and IV

    E. I, II, III and IV

72. The temperature of a liquid is raised at atmospheric pressure. Which property of liquids increases?

    A. Critical pressure

    B. Vapor pressure

    C. Surface tension

    D. Viscosity

    E. Boiling Point

NATURAL SCIENCES

73. Osmotic pressure is the pressure required to prevent _____ from flowing from low to high _____ concentration across a semipermeable membrane.

    A. solute, solute

    B. solute, solvent

    C. solvent, solute

    D. solvent, solvent

    E. ions, solute

74. Consider the reaction between iron and hydrogen chloride gas

    $$Fe(s) + 2HCl(g) \rightarrow FeCl_2(s) + H_2(g)$$

    7 moles of iron and 10 moles of HCl react until the limiting reagent is consumed. Which statements are true?

    I. HCl is the excess reagent
    II. HCl is the limiting reagent
    III. 7 moles of H_2 are produced
    IV. 2 moles of the excess reagent remain

    A. I and III

    B. I and IV

    C. II and III

    D. II and IV

    E. II, III and IV

75. 1-butanol, ethanol, methanol, and 1-propanol are all liquids at room temperature. Rank them in order of increasing viscosity.

    A. 1-butanol < 1-propanol < ethanol < methanol

    B. methanol < ethanol < 1-propanol < 1-butanol

    C. methanol < ethanol < 1-butanol < 1-propanol

    D. 1-propanol < 1-butanol < ethanol < methanol

    E. ethanol < methanol < 1-butanol < 1-propanol

76. List the following scientists in chronological order, from earliest to most recent, with respect to their most significant contribution to atomic theory

    I. John Dalton
    II. Niels Bohr
    III. J. J. Thomson
    IV. Ernest Rutherford

    A. I, III, II, IV

    B. I, III, IV, II

    C. I, IV, III, II

    D. III, I, II, IV

    E. I, II, III, IV

77. Which statement about acids and bases is NOT true?

    A. All strong acids ionize in water.

    B. All Lewis acids accept an electron pair.

    C. All Brønsted bases use OH⁻ as a proton acceptor.

    D. All Arrhenius acids form H^+ ions in water.

    E. The reaction of an acid with a base is called neutralization reaction.

78. Why does $CaCl_2$ have a higher normal melting point than NH_3?

    A. Covalent bonds are stronger than London dispersion forces.

    B. Covalent bonds are stronger than hydrogen bonds.

    C. Ionic bonds are stronger than London dispersion forces.

    D. Ionic bonds are stronger than hydrogen bonds.

    E. Covalent Bonds are stronger than Ionic Bonds.

79. Which intermolecular attraction explains the following trend in straight-chain alkanes?

Condensed structural formula	Boiling point (°C)
CH_4	-161.5
CH_3CH_3	-88.6
$CH_3CH_2CH_3$	-42.1
$CH_3CH_2CH_2CH_3$	-0.5
$CH_3CH_2CH_2CH_2CH_3$	36.0
$CH_3CH_2CH_2CH_2CH_2CH_3$	68.7

    A. London dispersion forces

    B. Dipole-dipole interactions

    C. Hydrogen bonding

    D. Ion-induced dipole interactions

    E. Covalent bonds

80. Which substance is most likely to be a gas at STP?

    A. SeO_2

    B. F_2

    C. $CaCl_2$

    D. I_2

    E. H_2O

NATURAL SCIENCES

81. **A calorie is the amount of heat energy that will**

    A. Raise the temperature of one gram of water from 14.5° C to 15.5° C.

    B. Lower the temperature of one gram of water from 16.5° C to 15.5° C

    C. Raise the temperature of one gram of water from 32° F to 33° F

    D. Cause water to boil at two atmospheres of pressure.

    E. Raise the temperature of 100 mL of water from 14.5° C to 15.5° C.

82. **Heat transfer by electromagnetic waves is termed**

    A. Conduction

    B. Convection

    C. Radiation

    D. Phase Change

    E. Warming

83. **What is temperature?**

    A. Temperature is a measure of the conductivity of the atoms or molecules in a material

    B. Temperature is a measure of the kinetic energy of the atoms or molecules in a material

    C. Temperature is a measure of the relativistic mass of the atoms or molecules in a material

    D. Temperature is a measure of the angular momentum of electrons in a material

    E. Temperature is the amount of heat of a material

84. **Which statement about reactions is true?**

    A. All spontaneous reactions are exothermic and cause an increase in entropy.

    B. An endothermic reaction that increases the order of the system cannot be spontaneous.

    C. A reaction can be non-spontaneous in one direction and also non-spontaneous in the opposite direction.

    D. Melting snow is an exothermic process.

    E. Thermodynamic Functions are dependent on the reaction pathway.

NATURAL SCIENCES

85. If the internal energy of a system remains constant, how much work is done by the system if 1 kJ of heat energy is added?

    A. 0 kJ

    B. -1 kJ

    C. 1 kJ

    D. 3.14 kJ

    E. 0.5 kJ

86. When KNO_3 dissolves in water, the water grows slightly colder. An increase in temperature will _____ the solubility of KNO_3.

    A. Increase

    B. Slightly decrease

    C. Have no effect on

    D. Have an unknown effect on

    E. Highly decrease

87. Which phase may be present at the triple point of a substance?

    I. Gas
    II. Liquid
    III. Solid
    IV. Supercritical fluid

    A. I, II, and III

    B. I, II, and IV

    C. II, III, and IV

    D. I, II, III, and IV

    E. I and III

88. The normal boiling point of water on the Kelvin scale is closest to

    A. 100 K

    B. 112 K

    C. 212 K

    D. 273 K

    E. 373 K

89. A few minutes after opening a bottle of perfume, the scent is detected on the other side of the room. What law relates to this phenomenon?

    A. Graham's law

    B. Dalton's law

    C. Boyle's law

    D. Avogadro's law

    E. Charles' law

NATURAL SCIENCES

90. **The magnitude of a force is**

    A. Directly proportional to mass and inversely to acceleration

    B. Inversely proportional to mass and directly to acceleration

    C. Directly proportional to both mass and acceleration

    D. Inversely proportional to both mass and acceleration

    E. Independent of the mass

91. **When acceleration is plotted versus time, the area under the graph represents**

    A. Moment in time

    B. Distance

    C. Velocity

    D. Acceleration

    E. Mass

92. **A needle floating in a tray of water demonstrates the property of**

    A. Specific Heat

    B. Surface Tension

    C. Oil-Water Interference

    D. Archimedes' Principle

    E. Metal density

93. **Heat is added to a pure solid at its melting point until it all becomes liquid at its freezing point. Which of the following occur?**

    I. Intermolecular attractions are weakened.
    II. The kinetic energy of the molecules does not change.
    III. The freedom of the molecules to move about increases.

    A. I only

    B. II only

    C. III only

    D. I, II and III

    E. I and III

94. **Which statement about thermochemistry is true?**

    A. Particles in a system move about less freely at high entropy.

    B. Water at 100° C has the same internal energy as water vapor at 100° C.

    C. A decrease in the order of a system corresponds to an increase in entropy.

    D. At its sublimation temperature, dry ice has higher entropy than gaseous CO_2.

    E. A decrease in the order of a system corresponds to a decrease in entropy.

95. A semi-conductor allows current to flow

    A. Never

    B. Always

    C. As long as it stays below a maximum temperature

    D. When a minimum voltage is applied

    E. Only when a high voltage is applied

96. What is the direction of the magnetic field at the center of the loop of current (I) shown below (i.e., at point A)?

    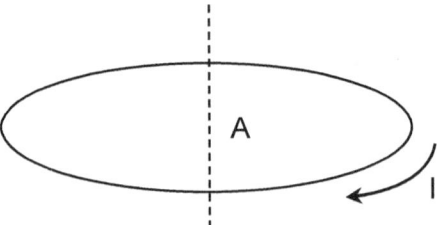

    A. Down, along the axis (dotted line)

    B. Up, along the axis (dotted line)

    C. The magnetic field is oriented in a radial direction

    D. The direction is random

    E. There is no magnetic field at point A

97. An electromagnetic wave propagates through a vacuum. Independent of its wavelength, it will move with constant

    A. Acceleration

    B. Velocity

    C. Induction

    D. Sound

    E. Frequency

98. A light bulb is connected in series with a rotating coil within a magnetic field. Which actions can increase the brightness of the light?

    I. Rotating the coil more rapidly
    II. Using more loops in the coil
    III. Using a different color wire for the coil
    IV. Using a stronger magnetic field

    A. I only

    B. II only

    C. I, II and IV

    D. I, II and III

    E. I, II, III and IV

NATURAL SCIENCES

99. Rainbows are created by

A. Reflection, dispersion, and recombination

B. Reflection, resistance, and expansion

C. Reflection, compression, and specific heat

D. Reflection, refraction, and dispersion

E. Reflection, refraction and recombination

100. The Sun is a

A. Asteroid

B. Star

C. Planet

D. Satellite

E. Giant planet

101. The nucleus of comets is made of

A. Gases only

B. Dust only

C. Solid rocks

D. Ice and gases

E. Ice, dust and rocky particles

102. What is the warmest planet of the Solar System?

A. Mars

B. Saturn

C. Earth

D. Venus

E. Jupiter

103. The mass of the Milky Way is mainly composed by

A. Dark matter

B. Stars

C. Planets

D. Dust

E. Asteroids

104. The light that reaches the planets originates in the

A. Dwarf planets

B. Black holes

C. Asteroids

D. Moons

E. Stars

NATURAL SCIENCES

105. In 2015, the *New Horizons* spacecraft flew by Pluto, a dwarf planet of the Solar System. How many moons does Pluto have?

    A. Pluto doesn't have moons.

    B. Two moons

    C. Three moons

    D. Four moons

    E. Five moons

106. Geostationary satellites are often used by communications. Their orbital period is ____ Earth's rotational period, and their direction of rotation is ____

    A. Shorter than – the same as Earth's

    B. Longer than – opposite to Earth's

    C. Equal to – the same as Earth's

    D. Shorter than – opposite to Earth's

    E. Longer than – the same as Earth's

107. Which list shows only terrestrial planets?

    A. Earth, Mercury, Jupiter, and Mars

    B. Earth, Mercury, Venus, and Mars

    C. Jupiter, Saturn, Neptune, and Uranus

    D. Earth, Mercury, Neptune, and Uranus

    E. Earth, Mercury, Venus, and Uranus

108. Why Mars is called "the red planet"?

    A. Its waters are rich in sulfur compounds

    B. Its waters are rich in iron compounds

    C. Due to gases present in the atmosphere

    D. Iron oxides in its surface give a reddish appearance

    E. Its surface is rich in sulfur compounds

109. Which layer of the atmosphere contains the ozone layer?

    A. Stratosphere

    B. Thermosphere

    C. Mesosphere

    D. Troposphere

    E. Ionosphere

110. What is the most abundant gas in our atmosphere?

    A. Oxygen

    B. Nitrogen

    C. Helium

    D. Argon

    E. Carbonic dioxide

NATURAL SCIENCES

111. **Which process adds oxygen to the atmosphere?**

    A. Respiration

    B. Ultraviolet light

    C. Weathering of rocks

    D. Photosynthesis

    E. Movements of air masses

112. **What is the impact of sulfur oxides and nitrogen oxides in the environment when they react with water?**

    A. Ammonia

    B. Acidic precipitation

    C. Carbonic acid

    D. Global warming

    E. Greenhouse effect

113. **Most of Earth's freshwater is**

    A. Available for human consumption

    B. Found in rivers and lakes

    C. Found in the groundwater

    D. Found in the clouds and atmosphere

    E. Trapped in the form of ice

114. **Which of the following statements about groundwater is true?**

    I. Groundwater is a renewable resource
    II. Groundwater cannot be replaced by sea water
    III. Groundwater never flows to the surface of the Earth

    A. I only

    B. II only

    C. III only

    D. I and II

    E. I, II and III

115. **What are the current Era, Period and Epoch we are living in, respectively?**

    A. Cenozoic, Quaternary, Holocene

    B. Mesozoic, Quaternary, Holocene

    C. Mezozoic, Quaternary, Pleistocene

    D. Cenozoic, Neogene, Holocene

    E. Cenozoic, Neogene, Eocene

116. **What is the main element present in the Earth's core?**

    A. Carbon

    B. Sulfur

    C. Iron

    D. Cobalt

    E. Manganese

NATURAL SCIENCES

117. The lava released by volcanoes during eruption originates in

   A. The Earth's inner mantle

   B. The volcano itself

   C. The Earth's outer core

   D. The Earth's inner core

   E. Magma chambers

118. Which statement about the carbon cycle is NOT true?

   A. Ten percent of all available carbon is in the air.

   B. Carbon dioxide is fixed by glycosylation.

   C. Plants fix carbon in the form of glucose.

   D. Animals release carbon through respiration.

   E. Most atmospheric carbon comes from the decay of dead organisms.

119. The supercontinent Pangea was fragmented due to which geological process?

   A. Sedimentation

   B. Erosion

   C. Chemical weathering

   D. Plate tectonics

   E. Meteorite impact

120. Which statement(s) about fossils is (are) true?

   I. Only animals with bones can become fossils
   II. There are insect fossils found in amber
   III. The fossil's age can be estimated by radiometric aging
   IV. The fossil's age can be estimated by stratigraphy

   A. I only

   B. II only

   C. I and II

   D. II, III and IV

   E. I, II, III and IV

NATURAL SCIENCES

ANSWER KEY

Question Number	Correct Answer	Your Answer	Question Number	Correct Answer	Your Answer	Question Number	Correct Answer	Your Answer
1	B		41	B		81	A	
2	D		42	E		82	C	
3	E		43	C		83	B	
4	A		44	D		84	B	
5	D		45	D		85	C	
6	D		46	B		86	A	
7	B		47	E		87	A	
8	C		48	C		88	E	
9	C		49	D		89	A	
10	C		50	E		90	C	
11	E		51	C		91	C	
12	C		52	B		92	B	
13	A		53	C		93	D	
14	A		54	B		94	C	
15	A		55	A		95	D	
16	C		56	E		96	A	
17	B		57	C		97	B	
18	D		58	D		98	C	
19	C		59	C		99	D	
20	E		60	A		100	B	
21	E		61	B		101	E	
22	A		62	E		102	D	
23	D		63	A		103	A	
24	A		64	C		104	E	
25	C		65	E		105	E	
26	B		66	B		106	C	
27	D		67	B		107	B	
28	B		68	D		108	D	
29	B		69	A		109	A	
30	E		70	C		110	B	
31	A		71	D		111	D	
32	D		72	B		112	B	
33	A		73	C		113	E	
34	E		74	D		114	D	
35	C		75	B		115	A	
36	A		76	B		116	C	
37	B		77	C		117	E	
38	C		78	D		118	B	
39	D		79	A		119	D	
40	A		80	B		120	D	

NATURAL SCIENCES

RATIONALES

1. **According to scientists, what is the estimate age of the Earth?**

    A. 4.5 million years

    B. 4.5 billion years

    C. 450 million years

    D. 1.000 million years

    E. 10.000 million years

The answer is B.
The estimated age of Earth, based on radiometric age dating, is 4.5 billion years.

2. **The first cells that evolved on earth were probably of which type?**

    A. Autotrophic

    B. Eukaryotic

    C. Similar to viruses

    D. Prokaryotic

    E. Endosymbiotic

The answer is D.
Prokaryotic cells are simpler than eukaryotic ones, and are thought to have originated first.

NATURAL SCIENCES

3. **What is a major principle of the Endosymbiotic Theory?**

    A. Birds and dinosaurs share a common ancestor.

    B. Animals evolved in close relationships with one another.

    C. Prokaryotes arose from eukaryotes.

    D. Inorganic compounds are the basis of living things.

    E. Eukaryotes arose from very simple prokaryotes.

The answer is E.
The Endosymbiotic theory of the origin of eukaryotes states that eukaryotes arose from symbiotic groups of prokaryotic cells. According to this theory, smaller prokaryotes lived within larger prokaryotic cells, eventually evolving into chloroplasts and mitochondria.

4. **According to Oparin & Haldane's theory, the primitive atmosphere was composed by**

    A. Hydrogen, methane, water, ammonia

    B. Oxygen, methane, water, ammonia

    C. Oxygen and carbonic gas

    D. Oxygen, carbonic gas, nitrogen

    E. Hydrogen, methane, water, ozone

The answer is A.
According to Oparin and Haldane, the primitive atmosphere was reducing and without free oxygen or ozone (O_3).

NATURAL SCIENCES

5. **Which of these is true about natural selection?**

   A. It acts on an individual genotype

   B. It is not currently happening

   C. It is only an animal phenomenon

   D. It acts on the individual phenotype

   E. It is used to prevent overpopulation

The answer is D.
Natural selection is a process that occurs in all living things, is currently happening, and acts on the phenotype. Natural selection cannot be used to prevent overpopulation.

6. **Which of these is a result of reproductive isolation?**

   A. Extinction

   B. Migration

   C. Fossilization

   D. Speciation

   E. Radiation

The answer is D.
Reproductive isolation is caused by any factor that impedes two species from producing viable, fertile hybrids. Reproductive isolation of populations is the primary criterion for recognition of species status.

NATURAL SCIENCES

7. Which of these is NOT a prezygotic barrier?

   A. Geographical isolation

   B. Hybrid sterility

   C. Temporal isolation

   D. Mechanical isolation

   E. Behavioral isolation

The answer is B.
Hybrid sterility occurs after fertilization, where the hybrid is incapable of producing viable gametes. The other alternatives—geographical, temporal, mechanical, and behavioral isolation—prevent reproduction and are termed prezygotic barriers.

8. Which mode of natural selection favors the more common phenotypes?

   A. Directional selection

   B. Positive selection

   C. Stabilizing selection

   D. Diversifying selection

   E. Disruptive selection

The answer is C.
Directional, or positive selection, changes the frequency of phenotypes in one direction. Disruptive, or diversifying selection, favors individuals on both extremes of the phenotypic range.

NATURAL SCIENCES

9. Which phylum accounts for 85% of all animal species?

    A. Nematoda

    B. Chordata

    C. Arthropoda

    D. Cnidaria

    E. Annelida

The answer is C.
The Arthropoda is a very rich phylum that includes hundreds of thousands species of insects, crustaceans, arachnids and others.

10. The scientific name of humans is *Homo sapiens*. Choose the proper classification beginning with kingdom and ending with order

    A. Animalia, Vertebrata, Mammalia, Primates, Hominidae

    B. Animalia, Vertebrata, Chordata, Mammalia, Primates

    C. Animalia, Chordata, Vertebrata, Mammalia, Primates

    D. Chordata, Vertebrata, Primate, *Homo, sapiens*

    E. Chordata, Primates, Hominidae, *Homo, sapiens*

The answer is C.
According to the current zoological classification, Animalia is a kingdom, Chordata is a phylum, Mammalia is a class, and Primates is an order. *Homo* is the genus name, and *sapiens* is the specific epithet. Hominidae is a family.

NATURAL SCIENCES

11. **Which of the following animals is coelomate?**

    I. Flatworms
    II. Earthworms
    III. Crickets

    A. I only

    B. II only

    C. III only

    D. I and III

    E. II and III

The answer is E.
Flatworms are acoelomates since they lack a defined body cavity. Earthworms and crickets have a true body cavity derived from the mesoderm, the coelom.

12. **Heterotrophic organisms that have cell walls with chitin are classified as**

    A. Plants

    B. Bacteria

    C. Fungi

    D. Animals

    E. Protists

The answer is C.
Protists and animals have no cell walls. The cell wall of plants is mainly composed of lignin and cellulose, whereas bacteria have walls composed of polysaccharides.

NATURAL SCIENCES

13. **Of what are viruses made?**

    A. A protein coat surrounding a nucleic acid

    B. RNA and protein surrounded by a cell wall

    C. A nucleic acid surrounding a protein coat

    D. Protein surrounded by DNA

    E. A lipid bilayer surrounding a protein coat and RNA

The answer is A.
Viruses are composed of a protein coat and a nucleic acid, which can be either RNA or DNA. Viruses have no lipid layers or cell walls.

14. **According to the fluid-mosaic model of the cell membrane, membranes are composed of**

    A. A phospholipid bilayer with proteins embedded in the layers

    B. One layer of phospholipids with cholesterol embedded in the layer

    C. Two layers of protein with lipids embedded in the layers

    D. DNA and fluid proteins

    E. Two layers of phospholipids and DNA

The answer is A
DNA is not present in the membrane, and the layers are formed by phospholipids, not proteins.

NATURAL SCIENCES

15. Which of the following is not part of the cytoskeleton?

    A. Vacuoles

    B. Microfilaments

    C. Microtubules

    D. Intermediate filaments

    E. Motor proteins

The answer is A

Vacuoles are organelles filled with water and surrounded by a membrane. They perform a variety of functions but are not related to the cytoskeleton.

16. Bacteria commonly reproduce by a process called binary fission. Which of the following best defines this process?

    A. Viral vectors carry DNA to new bacteria

    B. DNA from one bacterium enters another

    C. DNA doubles and the bacterial cell divides

    D. DNA from dead cells is absorbed into bacteria

    E. Bacteria merge with others to form new species

The answer is C

Binary fission is the asexual process in which the bacteria divide in half after the DNA doubles. This results in an exact genetic clone of the parent cell.

NATURAL SCIENCES

17. **What is the stage of mitosis shown in the diagram?**

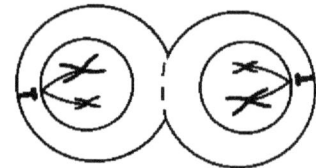

   A. Prophase

   B. Telophase

   C. Anaphase

   D. Metaphase

   E. Interphase

The answer is B

Telophase is the last stage of mitosis. At this stage, two nuclei become visible and the nuclear membrane reassembles.

18. **Crossing over, which increases genetic diversity, occurs during which stage of meiosis?**

   A. Telophase II in meiosis

   B. Metaphase in mitosis

   C. Interphase in meiosis

   D. Prophase I in meiosis

   E. Metaphase II in meiosis

The answer is D

During prophase I of meiosis, the replicated chromosomes condense and pair with their homologues in a process called synapsis. During this process there is an exchange of genetic material between the homologues.

NATURAL SCIENCES

19. What is the enzyme that unwinds DNA during replication?

    A. DNAse

    B. Replicase

    C. DNA helicase

    D. DNA topoisomerases

    E. DNA polymerase

The answer is C
A DNA polymerase synthesizes new DNA, and replicases are a special type of polymerases. Topoisomerases help relaxing the molecule supercoil, and DNases degrade the DNA.

20. Which part of a DNA nucleotide can vary?

    A. Deoxyribose

    B. Phosphate group

    C. Hydrogen bonds

    D. Sugar

    E. Nitrogenous base

The answer is E
DNA is a polymer formed by a sugar molecule, which is always a deoxyribose, and hydrogen bonds and phosphate groups are always present. The nitrogenous base, however, can be an adenine, cytosine, guanine, or thymine.

NATURAL SCIENCES

21. Which of these describes facilitated diffusion?

    A. It requires energy

    B. It only happens in plant cells

    C. It only allows molecules to leave a cell but not to enter it

    D. It produces a significant amount of energy for the cell

    E. It needs a transport molecule to pass through the membrane

The answer is E

Facilitated diffusion requires no energy but needs a transport molecule to pass a molecule through the membrane.

22. During which part of photosynthesis is oxygen given off?

    A. Light reactions

    B. Dark reactions

    C. Krebs cycle

    D. Reduction of NAD+ to NADH

    E. Phosphorylation

The answer is A.

The conversion of solar energy to chemical energy occurs in the light reactions. Electrons are transferred when chlorophyll absorbs light and causes water to split, releasing oxygen as a waste product.

NATURAL SCIENCES

23. During the Krebs cycle, 8 carrier molecules are formed. What are they?

A. 3 NADH, 3 FADH, 2 ATP

B. 6 NADH and 2 ATP

C. 4 FADH$_2$ and 4 ATP

D. 6 NADH and 2 FADH$_2$

E. 4 NADH and 4 FADH$_2$

The answer is D.
For each molecule of CoA that enters the Kreb's cycle, 3 NADH and 1 FADH$_2$ are formed. There are 2 molecules of CoA formed from each glucose, so the total yield is 6 NADH and 2 FADH$_2$ during the Krebs cycle.

24. In the comparison of respiration to photosynthesis, which statement is true?

A. Oxygen is a waste product in photosynthesis but not in respiration

B. Glucose is produced in respiration but not in photosynthesis

C. Carbon dioxide is formed in photosynthesis but not in respiration

D. Water is formed in respiration but not in photosynthesis

E. Carbon dioxide and water are formed in photosynthesis

The answer is A.
In photosynthesis, water is split and the oxygen is given off as a waste product. In respiration, water and carbon dioxide are the waste products. Glucose is produced by photosynthesis, while CO$_2$ is formed in respiration. Water is formed by respiration and photosynthesis.

NATURAL SCIENCES

25. Identify the correct sequence of organization of living things

    A. Cell – organelle – organ – tissue – organ system – organism

    B. Cell – tissue – organ – organelle – organ system – organism

    C. Organelle – cell – tissue – organ – organ system – organism

    D. Organ system – tissue – organelle – cell – organism – organ

    E. Organism – organ system – tissue – cell – organelle – organ

The answer is C.
An organism, such as a human, is composed of several organ systems such as the circulatory and nervous systems. These organ systems consist of many organs including the heart and the brain. Organs are made of tissue such as cardiac muscle. Tissues are made up of cells, which contain organelles like the mitochondria and the Golgi apparatus.

26. Bird wings, human arms, and the flipper of whales have the same bone structure and different functions. These are called

    A. Polymorphic structures

    B. Homologous structures

    C. Vestigial structures

    D. Analogous structures

    E. Primitive structure

The answer is B.
Homologous structures have the same genetic basis (leading to similar appearances), but are used for different functions.

NATURAL SCIENCES

27. **Parts of the nervous system include all but the following**

    A. Brain

    B. Spinal cord

    C. Axons

    D. Venules

    E. Glial cells

The answer is D.
Venules are small blood vessels, thus they are part of the circulatory system.

28. **Which of the following best completes the statement below?**
    Peristalsis and the movement of the iris are possible due to the action of _____ muscles.

    A. Skeletal

    B. Smooth

    C. Cardiac

    D. Striated

    E. Voluntary

The answer is B.
Smooth muscles are non-striated and responsible for involuntary movements, such as peristalsis and the iris movement. Skeletal muscles, on the other hand, are striate and responsible for voluntary movements. The cardiac muscle is found in the heart.

NATURAL SCIENCES

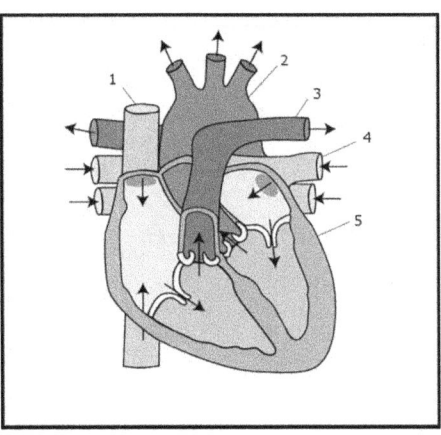

29. Consider the heart illustration above, with arrows indicating the direction of blood flow. Which number indicates the aorta?

   A. 1

   B. 2

   C. 3

   D. 4

   E. 5

The answer is B.
Number 1 indicates the vena cava, number 3 is the pulmonary artery, number 4 is the pulmonary vein, and number 5 is the cardiac muscle.

30. Microorganisms use all but which of the following for locomotion?

   A. Pseudopods

   B. Flagella

   C. Cilia

   D. Pili

   E. Villi

The answer is E.
Villi are protusions of the intestine.

NATURAL SCIENCES

31. Which of these is a function of the cardiovascular system?

    A. Move oxygenated blood around the body

    B. Oxygenate the blood through gas exchange

    C. Act as an exocrine system

    D. Flush toxins out of the body

    E. Transport signals from the brain

The answer is A.
The cardiovascular system moves oxygenated blood around the body via the heart (a pump) and tubes (arteries and veins).

32. Which of these steroids is not created in the gonads?

    A. Testosterone

    B. Estrogen

    C. Progesterone

    D. ACTH

    E. FSH

The answer is D.
ACTH, or Adrenocorticotropic hormone, is produced by the pituitary gland.

33. The role of neurotransmitters in nerve action is

    A. To turn off the sodium pump

    B. To turn off the calcium pump

    C. To send impulses to neurons

    D. To send impulses to the body

    E. To maintain the membrane polarized

The answer is A.
The neurotransmitters turn off the sodium pump, which results in depolarization of the membrane.

NATURAL SCIENCES

34. Which of the following is NOT a function of the vertebrate skin

    A. Respiration

    B. Protection

    C. Sensation

    D. Regulation of temperature

    E. Regulation of reproduction

The answer is E.
The skin performs a variety of functions but is not involved in reproduction. It serves as a protective barrier against infection. It contains hair follicles that respond to sensation, and it plays a role in thermoregulation. Also, it can store lipids and it mediates respiration in certain animals.

35. Homeostatic mechanisms in the body do NOT include

    A. Thermoregulation

    B. Excretion

    C. Respiration

    D. Osmoregulation

    E. Hemostasis

The answer is C.
All but respiration are homeostatic mechanisms used by the body to achieve homeostasis.

NATURAL SCIENCES

36. **What controls gas exchange on the bottom of a plant leaf?**

    A. Stomata

    B. Epidermis

    C. Collenchyma and schlerenchyma

    D. Palisade mesophyll

    E. Trichomes

The answer is A.
Stomata provide openings on the underside of leaves for oxygen to move in or out of the plant and for carbon dioxide to move in.

37. **Hormones are essential to the regulation of reproduction. What organ is responsible for the release of hormones for sexual maturity?**

    A. Thymus gland

    B. Hypothalamus

    C. Pancreas

    D. Thyroid gland

    E. Cerebellum

The answer is B.
The hypothalamus begins secreting hormones that help mature the reproductive system and stimulate development of the secondary sex characteristics.

NATURAL SCIENCES

38. Fertilization in humans usually occurs in the

    A. Uterus

    B. Ovary

    C. Fallopian tubes

    D. Vagina

    E. Cervix

The answer is C.
Fertilization of the egg by the sperm normally occurs in a fallopian tube. The fertilized egg is then implanted in the uterine lining for development.

39. After sea turtles hatch on the beach, they start the journey to the ocean. This is due to

    A. Learned behavior

    B. Territoriality

    C. The tide

    D. Innate behavior

    E. Feeding strategy

The answer is D.
Innate behavior is inborn or instinctual. The baby sea turtles did not learn from their mother. They immediately knew to head toward the sea once they hatched.

NATURAL SCIENCES

40. What is any foreign particle called that causes an immune reaction?

A. An antigen

B. A histocompatibility complex

C. An antibody

D. A vaccine

E. A bacteriophage

The answer is A.
Histocompatibility complex is a set of molecules produced by organisms. Antibodies are produced by the body, and vaccines are substances that stimulate their production. Bacteriophages are viruses that infect bacteria.

41. What is (are) the germ layer(s) missing in diploblastic animals?

A. Ectoderm only

B. Mesoderm only

C. Endoderm only

D. Ectoderm and mesoderm

E. Endoderm and mesoderm

The answer is B.
Diploblastic animals have only two germ layers, the ectoderm and the endoderm. In comparison with triploblastic animas, they lack the mesoderm.

NATURAL SCIENCES

42. What is the order of the stages that happen after fertilization?

A. Blastula – gastrulation – neurulation – organogenesis – cleavage

B. Cleavage – neurulation – gastrulation – organogenesis – blastula

C. Cleavage – blastula – gastrulation – neurulation – cell growth

D. Cell growth – gastrulation – blastula – neurulation – organogenesis

E. Cleavage – blastula – gastrulation – neurulation – organogenesis

The answer is E.
Cleavage occurs after fertilization and increases the embryo's cell number and forms the blastula. During gastrulation the germ layers are formed, and in neurulation the neural tube is formed. The last stage is organogenesis, where cells differentiate to perform a variety of functions.

43. What is the general term for a change that affects the sequence of bases in a gene?

A. Deletion

B. Polyploidy

C. Mutation

D. Duplication

E. Substitution

The answer is C.
A mutation is an inheritable change in DNA. It may be an error in replication or a spontaneous rearrangement of one or more segments of DNA. Deletion and duplication are types of mutations. Polyploidy occurs when an organism has more than two complete chromosome sets.

NATURAL SCIENCES

44. What can be said about homozygous individuals?

A. They have two different alleles

B. They are of the same species

C. They exhibit the same features

D. They have a pair of identical alleles

E. They produce identical offspring

The answer is D.
Homozygous individuals have a pair of identical alleles while heterozygous individuals have two different alleles.

45. In a Punnett square with a single trait, what are the ratios of genotypes produced between two heterozygous individuals?

A. 1:2:2

B. 2:1:1

C. 1:1:1

D. 1:2:1

E. 2:2:2

The answer is D.
According to Mendel's laws, crosses between two heterozygous individuals for a single trait will produce 25% homozygous dominant offspring, 50% heterozygous offspring, and 25% homozygous recessive offspring, thus 1:2:1.

NATURAL SCIENCES

46. A child with type O blood has a father with type A blood and a mother with type B blood. The genotypes of the parents respectively would be which of the following?

    A. AA and BO

    B. AO and BO

    C. AA and BB

    D. AO and OO

    E. OO and BO

The answer is B.
Type O blood has 2 recessive O genes. A child receives one allele from each parent; therefore, each parent in this example must have an O allele. The father has type A blood with a genotype of AO and the mother has type B blood with a genotype of BO.

47. Which of these defines the Law of Segregation defined by Gregor Mendel?

    A. After meiosis, each new cell will contain an allele that is recessive.

    B. Only one of two alleles is expressed in a heterozygous organism.

    C. The allele expressed is always the dominant allele.

    D. Alleles of one trait do not affect the inheritance of alleles on another chromosome.

    E. When sex cells form, the two alleles that determine a trait will end up on different gametes.

The answer is E.
The Law of Segregation states that only one of the two possible alleles from each parent is passed on to the offspring.

NATURAL SCIENCES

48. Hemophilia and color-blindness are examples of

   A. Lethal alleles

   B. Codominance system

   C. Sex-linked traits

   D. Incomplete dominance

   E. Nondisjunction

The answer is C.
Recessive traits present in the Y chromosome are more often expressed in men, as is the case with hemophilia and color blindness.

49. Which of the following is NOT an abiotic factor?

   A. Temperature

   B. Rainfall

   C. Soil quality

   D. Predation

   E. Wind speed

The answer is D.
Living organisms prey on each other; therefore it is a biotic factor.

NATURAL SCIENCES

50. An experiment was performed to measure the growth of bacteria at different temperatures. The cultures were kept on a 12 hour light/dark cycle and given the same amount of nutrients. Which of these is the independent variable?

A. Growth of number of colonies

B. Amount of nutrients

C. Type of bacteria used

D. Light duration

E. Temperature

The answer is E.
The number of colonies is the dependent variable. The light duration and amount of nutrients were kept constant. The temperature was manipulated, so it was the independent variable.

51. Which term is not associated with the water cycle?

A. Precipitation

B. Transpiration

C. Fixation

D. Evaporation

E. Infiltration

The answer is C.
Water is recycled through the processes of evaporation and precipitation. Transpiration is the evaporation of water from leaves. Infiltration carries water away from the surface. Fixation is not associated with the water cycle.

NATURAL SCIENCES

52. Which trophic level has the highest ecological efficiency?

   A. Decomposers

   B. Producers

   C. Tertiary consumers

   D. Secondary consumers

   E. Primary consumers

The answer is B.
The amount of energy that is transferred between trophic levels is called the ecological efficiency. The visual of this is represented in a pyramid of productivity. The producers have the greatest amount of energy and are at the bottom of this pyramid.

53. What is NOT true about competition?

   A. May occur between very different species

   B. It is usually asymmetric, affecting one species more than the other

   C. It increases the amount of available resources

   D. May affect the abundance of competitors

   E. Competition is a common process in natural communities

The answer is C.
Competition decreases the amount of resources that are available to competitors.

NATURAL SCIENCES

54. Which of the following is true about parasites?

   I. All parasites are facultative
   II. Parasites can be either ecto- or endoparasites
   III. Parasites increase their hosts' fitness
   IV. Parasites always kill their hosts

   A. I only

   B. II only

   C. I and II

   D. I, II and III

   E. I, II, III and IV

The answer is B.
Parasites can be obligate or facultative. They usually decrease their hosts' fitness, but it is not common for parasites to kill their hosts.

55. A clownfish is protected by a sea anemone's tentacles, and in turn, the anemone receives uneaten food from the clownfish. What type of symbiosis is exemplified by this example?

   A. Mutualism

   B. Parasitism

   C. Commensalism

   D. Competition

   E. Amensalism

The answer is A.
Neither the clownfish nor the anemone cause harmful effects towards one another and they both benefit from their relationship. Mutualism occurs when two species that occupy a similar space benefit from their relationship.

NATURAL SCIENCES

56. **Since the industrial revolution, the size of the human population has been**

    A. Decreasing

    B. Stable

    C. Increasing slowly

    D. Changing randomly

    E. Increasing exponentially

The answer is E.
Demographic records show that the human population has been increasing exponentially since the 1800s.

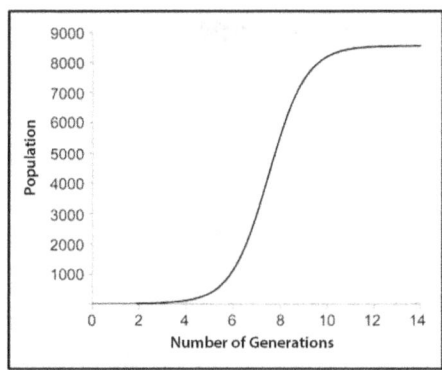

57. **According to the graphic above, what is the approximate population size after 8 generations?**

    A. 1000

    B. 2000

    C. 6000

    D. 8000

    E. 12000

The answer is C.
Following the curve along the x axis, at the stage of 8 generations, the y axis indicates the population size to be about 6000.

NATURAL SCIENCES

58. Xerophytes are more commonly found in which biome?

    A. Temperate forests

    B. Tropical forests

    C. Oceans

    D. Deserts

    E. Grasslands

The answer is D.
Xerophytes are plants with special adaptations to live in environments where there is a limited supply of liquid water, such as deserts.

59. Which zone of the ocean receives no light and has the lowest temperatures?

    A. Intertidal

    B. Pelagic

    C. Abyssal

    D. Euphotic

    E. Epipelagic

The answer is C.
The abyssal zone is located at about 4000 to 6000 meters depth. It never receives light and temperatures are 2°C to 3°C.

NATURAL SCIENCES

60. Which biome is the most prevalent on Earth?

    A. Marine

    B. Desert

    C. Savanna

    D. Tundra

    E. Lakes and rivers

The answer is A.
Oceans cover about 71% of the earth's surface

61. How many neutrons are there in $^{60}_{27}Co$?

    A. 27

    B. 33

    C. 60

    D. 87

    E. 14

The answer is B.
The number of neutrons is found by subtracting the atomic number (27) from the mass number (60).

NATURAL SCIENCES

62. Which of the following can be determined from the periodic table?

I. The number of protons
II. The number of neutrons
III. The number of isotopes of that atom
IV. The number of valence electrons

A. I only

B. I and II

C. I, II, and III

D. I, II, III, IV

E. I, II and IV

The answer is E.
The number of isotopes of an atom is not given in the periodic table.

63. When a radioactive material emits an alpha particle only, its atomic number

A. Decreases

B. Increases

C. Remains unchanged

D. Changes randomly

E. Approaches zero

The answer is A.
In alpha decay, a nucleus emits the equivalent of a helium atom. This includes two protons, so the original material changes its atomic number by a decrease of two.

NATURAL SCIENCES

64. The terrestrial composition of an element is 50.7% as a stable isotope with an atomic mass of 78.9 u and 49.3% as a stable isotope with an atomic mass of 80.9 u. What is the atomic mass of the element?

   A. 79.0 u

   B. 79.8 u

   C. 79.9 u

   D. 80.8 u

   E. 80.0 u

The answer is C.
This is calculated with the following equation:
Atomic mass of element = (Fraction of 1st isotope) (Atomic mass of 1st isotope) +
   (Fraction of 2nd isotope) (Atomic mass of 2nd isotope)
   = (0.507) (78.9 u) + (0.493) (80.9 u) + 79.89 u + 79.9 u

65. Moving down a column on the Periodic Table

   I. The atomic radius increase
   II. Ionization energy increase
   III. Protons are added
   IV. Metallic characteristics increase

   A. I only

   B. III only

   C I and II

   D. III and IV

   E. I, II, III

The answer is E.
In the periodic table, the metallic characteristics decrease from left to right.

NATURAL SCIENCES

66. Which of the following quantum numbers are needed to define the position of the electrons in an element?

  A. Principal, circular, magnetic and electromagnetic

  B. Principal, angular momentum, magnetic, and spin

  C. Angular, magnetic, electronic, spin

  D. Primary, angular momentum, magnetic, and spin

  E. Principal only

The answer is B.
Four quantum numbers describe an electron in an atom. Electronic and electromagnetic are not quantum numbers.

67. $^{3}_{1}H$ **decays with a half-life of 12 years. Twenty four years ago, 3.0 g of pure** $^{3}_{1}H$ **were placed in sealed container. How many grams of** $^{3}_{1}H$ **remain?**

  A. 0.38 g

  B. 0.75 g

  C. 1.5 g

  D. 0.125 g

  E. 3.0 g

The answer is B.
After 12 years, i.e. the half life, 50% of the original remain (1.50 g). After 12 years more, only 50% of this amount remain, thus 0.75 g.

NATURAL SCIENCES

68. Write a balanced nuclear equation for the emission of an alpha particle by polonium-209

  A. $^{209}_{84}\text{Po} \rightarrow {}^{205}_{81}\text{Po} + {}^{4}_{2}\text{He}$

  B. $^{209}_{84}\text{Po} \rightarrow {}^{205}_{82}\text{Po} + {}^{4}_{2}\text{He}$

  C. $^{209}_{84}\text{Po} \rightarrow {}^{209}_{85}\text{Po} + {}^{0}_{-1}\text{He}$

  D. $^{209}_{84}\text{Po} \rightarrow {}^{205}_{82}\text{Po} + {}^{4}_{2}\text{He}$

  E. $^{209}_{84}\text{Po} \rightarrow {}^{209}_{83}\text{Po} + {}^{4}_{2}\text{He}$

The answer is D.
The periodic table shows that polonium has an atomic number of 84. The emission of an alpha particle, which is ${}^{4}_{2}\text{He}$, will leave an atom with an atomic number of 82 and a mass number of 205.

69. Based on trends in the periodic table, which of the following properties would you expect to be greater for Rb than for K?

  I. Density
  II. Melting point
  III. Ionization energy
  IV. Oxidation number in a compound with chlorine

  A. I only

  B. I, II, and III

  C. II and III

  D. I and IV

  E. I, II, III, and IV

The answer is A.
Rb is underneath K in the alkali metal column (group 1) of the periodic table. There is a general trend for density to increase lower on the table for elements in the same row. Rb and K experience metallic bonds for intermolecular forces, and the strength of metallic bonds decreases for larger atoms further down the periodic table resulting in a lower melting point for Rb. Ionization energy decreases for larger atoms further down the periodic table. Both Rb and K would be expected to have a charge of +1 and therefore an oxidation number of +1 in a compound with chlorine.

NATURAL SCIENCES

70. According to the periodic table, which list contains only metals?

    A. Li, Cd, Ca, S

    B. Li, Cd, Ca, He

    C. Li, Cd, Ca, Pb

    D. Li, C, Ca, He

    F. Ca, Fe, F, Pb

The answer is C.
Sulfur (S), Carbon (C), and Fluorine (F) are a non-metals, Helium (He) is a noble gas. Li, Cd, Ca and Pb are metals.

71. Which of the following are the properties of noble gases?

    I. They are colorless and odorless under STP
    II. They have little tendency to gain or lose electrons
    III. High melting points
    IV. Full valence electron shells

    A. I and II

    B. I and III

    C. II and III

    D. I, II and IV

    E. I, II, III and IV

The answer is D.
Noble gases are colorless and odorless under STP. As they have full valence electron shells, they have little tendency to gain or lose electrons. The have low melting points and are gases at room temperatures.

NATURAL SCIENCES

72. **The temperature of a liquid is raised at atmospheric pressure. Which property of liquids increases?**

    A. Critical pressure

    B. Vapor pressure

    C. Surface tension

    D. Viscosity

    E. Boiling Point

The answer is B.
The critical pressure of a liquid is its vapor pressure at the critical temperature, and is always a constant value. A rising temperature increases the kinetic energy of molecules and decreases the importance of intermolecular attraction. More molecules will be free to escape to the vapor phase (vapor pressure increases), but the effect of attractions at the liquid-gas interface will fall (surface tension decreases) and molecules will flow against each other more easily (viscosity decreases).

73. **Osmotic pressure is the pressure required to prevent _____ from flowing from low to high _____ concentration across a semipermeable membrane.**

    A. solute, solute

    B. solute, solvent

    C. solvent, solute

    D. solvent, solvent

    E. ions, solute

The answer is C.
Osmotic pressure is the pressure required to prevent osmosis, which is the flow of solvent across the membrane from low to high solute concentration. This is also the direction from high to low solvent concentration.

NATURAL SCIENCES

74. Consider the reaction between iron and hydrogen chloride gas

$$Fe(s) + 2HCl(g) \rightarrow FeCl_2(s) + H_2(g)$$

7 moles of iron and 10 moles of HCl react until the limiting reagent is consumed. Which statements are true?

I. HCl is the excess reagent
II. HCl is the limiting reagent
III. 7 moles of H_2 are produced
IV. 2 moles of the excess reagent remain

A. I and III

B. I and IV

C. II and III

D. II and IV

E. II, III and IV

The answer is D.
The limiting reagent is found by dividing the number of moles of each reactant by its stoichiometric coefficient. The lowest result is the limiting reagent:

7 mol Fe x (1 mol reaction/1 mol Fe) = 7 mol reaction if Fe is limiting

10 mol HCl x (1 mol reaction/2 mol HCl) = 5 mol reaction if HCl is limiting.

Therefore, HCl is the limiting reagent (II is true) and Fe is the excess reagent. As 5 moles of the reaction take place, 5 moles of H_2 are produced, and of the 7 moles of Fe supplied, 5 are consumed, leaving 2 moles of the excess reagent.

NATURAL SCIENCES

75. **1-butanol, ethanol, methanol, and 1-propanol are all liquids at room temperature. Rank them in order of increasing viscosity.**

    A. 1-butanol < 1-propanol < ethanol < methanol

    B. methanol < ethanol < 1-propanol < 1-butanol

    C. methanol < ethanol < 1-butanol < 1-propanol

    D. 1-propanol < 1-butanol < ethanol < methanol

    E. ethanol < methanol < 1-butanol<1-propanol

The answer is B.
Higher viscosities result from stronger intermolecular attractive forces. The molecules listed are all alcohols with the -OH functional group attached to the end of a straight-chain alkane. In other words, they all have the formula $CH_3(CH_2)_{n-1}OH$. The only difference between the molecules is the length of the alkane corresponding to the value of n. With all else identical, larger molecules have greater intermolecular attractive forces due to a greater molecular surface for the attractions.

76. **List the following scientists in chronological order, from earliest to most recent, with respect to their most significant contribution to atomic theory**

    I. John Dalton
    II. Niels Bohr
    III. J. J. Thomson
    IV. Ernest Rutherford

    A. I, III, II, IV

    B. I, III, IV, II

    C. I, IV, III, II

    D. III, I, II, IV

    E. I, II, III, IV

The answer is B.
Dalton founded modern atomic theory. J.J. Thomson determined that the electron is a subatomic particle, but he placed it in the center of the atom. Rutherford discovered that electrons surround a small dense nucleus. Bohr determined that electrons may only occupy discrete positions around the nucleus.

NATURAL SCIENCES

77. Which statement about acids and bases is NOT true?

    A. All strong acids ionize in water.

    B. All Lewis acids accept an electron pair.

    C. All Brønsted bases use OH^- as a proton acceptor.

    D. All Arrhenius acids form H^+ ions in water.

    E. The reaction of an acid with a base is called neutralization reaction.

The answer is C.
Choice A is the definition of a strong acid, Choice B is the definition of a Lewis acid, and Choice D is the definition of an Arrhenius acid. By definition, all Arrhenius bases form OH^- ions in water, and all Brønsted bases are proton acceptors. But not all Brønsted bases use OH^- as a proton acceptor. For example, NH_3 is a Brønsted base.

78. Why does $CaCl_2$ have a higher normal melting point than NH_3?

    A. Covalent bonds are stronger than London dispersion forces.

    B. Covalent bonds are stronger than hydrogen bonds.

    C. Ionic bonds are stronger than London dispersion forces.

    D. Ionic bonds are stronger than hydrogen bonds.

    E. Covalent Bonds are stronger than Ionic Bonds.

The answer is D.
A higher melting point will result from stronger intermolecular bonds. $CaCl_2$ is an ionic solid, whereas the dominant attractive forces between NH_3 molecules are hydrogen bonds. Also, London dispersion forces are weaker than covalent bonds.

79. Which intermolecular attraction explains the following trend in straight-chain alkanes?

Condensed structural formula	Boiling point (°C)
CH_4	-161.5
CH_3CH_3	-88.6
$CH_3CH_2CH_3$	-42.1
$CH_3CH_2CH_2CH_3$	-0.5
$CH_3CH_2CH_2CH_2CH_3$	36.0
$CH_3CH_2CH_2CH_2CH_2CH_3$	68.7

A. London dispersion forces

B. Dipole-dipole interactions

C. Hydrogen bonding

D. Ion-induced dipole interactions

E. Covalent bonds

The answer is A.
Alkanes are composed entirely of non-polar C-C and C-H bonds, resulting in no dipole interactions or hydrogen bonding. London dispersion forces increase with the size of the molecule, resulting in a higher temperature requirement to break these bonds and a higher boiling point.

80. Which substance is most likely to be a gas at STP?

A. SeO_2

B. F_2

C. $CaCl_2$

D. I_2

E. H_2O

The answer is B.
A gas at STP has a normal boiling point under 0°C. The substance with the lowest boiling point will have the weakest intermolecular attractive forces and will be the most likely gas at STP. F_2 has the lowest molecular weight, is not a salt, metal, or covalent network solid, and is nonpolar, indicating the weakest intermolecular attractive forces of the four choices. F_2 actually is a gas at STP, and the other three are solids.

NATURAL SCIENCES

81. A calorie is the amount of heat energy that will

    A. Raise the temperature of one gram of water from 14.5° C to 15.5° C.

    B. Lower the temperature of one gram of water from 16.5° C to 15.5° C

    C. Raise the temperature of one gram of water from 32° F to 33° F

    D. Cause water to boil at two atmospheres of pressure.

    E. Raise the temperature of 100 mL of water from 14.5° C to 15.5° C.

The answer is A.
The definition of a calorie is "the amount of energy to raise one gram of water by one degree Celsius."

82. Heat transfer by electromagnetic waves is termed

    A. Conduction

    B. Convection

    C. Radiation

    D. Phase Change

    E. Warming

The answer is C.
Radiation is the transfer of heat via electromagnetic waves (and can occur in a vacuum). Conduction is the transfer of heat through direct physical contact and molecules moving and hitting each other. Convection is the transfer of heat via density differences and flow of fluids. Phase change causes transfer of heat (though not of temperature) in order for the molecules to take their new phase.

NATURAL SCIENCES

83. What is temperature?

    A. Temperature is a measure of the conductivity of the atoms or molecules in a material

    B. Temperature is a measure of the kinetic energy of the atoms or molecules in a material

    C. Temperature is a measure of the relativistic mass of the atoms or molecules in a material

    D. Temperature is a measure of the angular momentum of electrons in a material

    E. Temperature is the amount of heat of a material

The answer is B.
Temperature is, in fact, a measure of the kinetic energy of the constituent components of a material. Thus, as a material is heated, the atoms or molecules that compose it acquire greater energy of motion. This increased motion results in the breaking of chemical bonds and in an increase in disorder, thus leading to melting or vaporizing of the material at sufficiently high temperatures.

84. Which statement about reactions is true?

    A. All spontaneous reactions are exothermic and cause an increase in entropy.

    B. An endothermic reaction that increases the order of the system cannot be spontaneous.

    C. A reaction can be non-spontaneous in one direction and also non-spontaneous in the opposite direction.

    D. Melting snow is an exothermic process.

    E. Thermodynamic Functions are dependent on the reaction pathway.

The answer is B.
All reactions that are both exothermic and cause an increase in entropy will be spontaneous, but the converse is not true. Some spontaneous reactions are exothermic but decrease entropy and some are endothermic and increase entropy. The reverse reaction of a non-spontaneous reaction will be spontaneous. Melting snow requires heat, so it is an endothermic process.

NATURAL SCIENCES

85. **If the internal energy of a system remains constant, how much work is done by the system if 1 kJ of heat energy is added?**

    A. 0 kJ

    B. -1 kJ

    C. 1 kJ

    D. 3.14 kJ

    E. 0.5 kJ

The answer is C.
According to the first law of thermodynamics, if the internal energy of a system remains constant, then any heat energy added to the system must be balanced by the system performing work on its surroundings. In the case of an ideal gas, the gas would necessarily expand when heated, assuming a constant internal energy was somehow maintained.

86. **When KNO_3 dissolves in water, the water grows slightly colder. An increase in temperature will _____ the solubility of KNO_3.**

    A. Increase

    B. Slightly decrease

    C. Have no effect on

    D. Have an unknown effect on

    E. Highly decrease

The answer is A.
The decline in water temperature indicates that the net solution process is endothermic (requiring heat). A temperature increase supplying more heat will favor the solution and increase solubility according to Le Chatelier's principle.

87. Which phase may be present at the triple point of a substance?

I. Gas
II. Liquid
III. Solid
IV. Supercritical fluid

A. I, II, and III

B. I, II, and IV

C. II, III, and IV

D. I, II, III, and IV

E. I and III

The answer is A.
Gas, liquid, and solid may exist together at the triple point.

88. The normal boiling point of water on the Kelvin scale is closest to

A. 100 K

B. 112 K

C. 212 K

D. 273 K

E. 373 K

The answer is E.
Temperature in Kelvin is equal to the temperature in Celsius plus 273.15. Since the normal boiling point of water is 100° C, it will boil at 373.15 K.

NATURAL SCIENCES

89. A few minutes after opening a bottle of perfume, the scent is detected on the other side of the room. What law relates to this phenomenon?

    A. Graham's law

    B. Dalton's law

    C. Boyle's law

    D. Avogadro's law

    E. Charles' law

The answer is A.
Graham's law describes the rate of diffusion (or effusion) of a gas, in this instance, the rate of diffusion of molecules in perfume vapor.

90. The magnitude of a force is

    A. Directly proportional to mass and inversely to acceleration

    B. Inversely proportional to mass and directly to acceleration

    C. Directly proportional to both mass and acceleration

    D. Inversely proportional to both mass and acceleration

    E. Independent of the mass

The answer is C.
According to Newton's 2nd Law, the net force is equal to mass times acceleration.

NATURAL SCIENCES

91. When acceleration is plotted versus time, the area under the graph represents

    A. Moment in time

    B. Distance

    C. Velocity

    D. Acceleration

    E. Mass

The answer is C.
The area under a graph will have units equal to the product of the units of the two axes.
Therefore, multiply units of acceleration by units of time: (length/time^2)(time).
This equals length/time, i.e. units of velocity.

92. A needle floating in a tray of water demonstrates the property of

    A. Specific Heat

    B. Surface Tension

    C. Oil-Water Interference

    D. Archimedes' Principle

    E. Metal density

The answer is B.
The needle floats on the water because although the needle is denser than the water, the surface tension of the water causes sufficient resistance to support the small needle.

NATURAL SCIENCES

93. Heat is added to a pure solid at its melting point until it all becomes liquid at its freezing point. Which of the following occur?

    I. Intermolecular attractions are weakened.
    II. The kinetic energy of the molecules does not change.
    III. The freedom of the molecules to move about increases.

    A. I only

    B. II only

    C. III only

    D. I, II and III

    E. I and III

The answer is D.
Intermolecular attractions are lessened during melting. This permits molecules to move about more freely, but there is no change in the kinetic energy of the molecules because the temperature has remained the same.

94. Which statement about thermochemistry is true?

    A. Particles in a system move about less freely at high entropy.

    B. Water at 100° C has the same internal energy as water vapor at 100° C.

    C. A decrease in the order of a system corresponds to an increase in entropy.

    D. At its sublimation temperature, dry ice has higher entropy than gaseous CO_2.

    E. A decrease in the order of a system corresponds to a decrease in entropy.

The answer is C.
Entropy may be thought of as the disorder in a system. At high entropy, particles have a large freedom of molecular motion. Water and water vapor at 100° C contain the same translational kinetic energy, but water vapor has additional internal energy in the form of resisting the intermolecular attractions between molecules. Water vapor has a higher internal energy because heat must be added to boil water. Sublimation is the phase change from solid to gas, and there is less freedom of motion for particles in solids than in gases. Solid CO_2 (dry ice) has a lower entropy than gaseous CO_2 because entropy decreases during a phase change that prevents molecular motion.

NATURAL SCIENCES

95. A semi-conductor allows current to flow

A. Never

B. Always

C. As long as it stays below a maximum temperature

D. When a minimum voltage is applied

E. Only when a high voltage is applied

The answer is D.
Semiconductors do not conduct as well as conductors, but they conduct better than insulators. Semiconductors can conduct better when the temperature is higher, and their electrons move most readily under a potential difference.

96. What is the direction of the magnetic field at the center of the loop of current (I) shown below (i.e., at point A)?

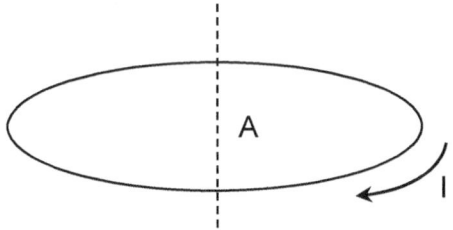

A. Down, along the axis (dotted line)

B. Up, along the axis (dotted line)

C. The magnetic field is oriented in a radial direction

D. The direction is random

E. There is no magnetic field at point A

The answer is A.
The magnetic field may be found by applying the right-hand rule. The magnetic field curls around the wire in the direction of the curled fingers when the thumb is pointed in the direction of the current. Since there is a degree of symmetry, with point A lying in the center of the loop, the contributions of all the current elements on the loop must yield a field that is either directed up or down at the axis. Use of the right-hand rule indicates that the field is directed down.

NATURAL SCIENCES

97. An electromagnetic wave propagates through a vacuum. Independent of its wavelength, it will move with constant

   A. Acceleration

   B. Velocity

   C. Induction

   D. Sound

   E. Frequency

The answer is B.
Electromagnetic waves are considered always to travel at the speed of light.

98. A light bulb is connected in series with a rotating coil within a magnetic field. Which actions can increase the brightness of the light?

   I. Rotating the coil more rapidly
   II. Using more loops in the coil
   III. Using a different color wire for the coil
   IV. Using a stronger magnetic field

   A. I only

   B. II only

   C. I, II and IV

   D. I, II and III

   E. I, II, III and IV

The answer is C.
A rotating coil in a magnetic field generates electric current, according to Faraday's Law. Faraday's Law states that the amount of electromotive force generated is proportional to the rate of change of magnetic flux through the loop. Force increases if the coil is rotated more rapidly, if there are more loops, or if the magnetic field is stronger.

NATURAL SCIENCES

99. Rainbows are created by

    A. Reflection, dispersion, and recombination

    B. Reflection, resistance, and expansion

    C. Reflection, compression, and specific heat

    D. Reflection, refraction, and dispersion

    E. Reflection, refraction and recombination

The answer is D.
Rainbows are formed by light that goes through water droplets and is dispersed into its colors. Refraction is important in bending the differently colored light waves.

100. The Sun is a

    A. Asteroid

    B. Star

    C. Planet

    D. Satellite

    E. Giant planet

The answer is B.
The sun is the star of the solar system.

101. The nucleus of comets is made of

    A. Gases only

    B. Dust only

    C. Solid rocks

    D. Ice and gases

    E. Ice, dust and rocky particles

The answer is E.
Comets are small bodies of the solar system. They are made of ice with nuclei made of loose ice, dust, and rocky particles.

NATURAL SCIENCES

102. What is the warmest planet of the Solar System?

    A. Mars

    B. Saturn

    C. Earth

    D. Venus

    E. Jupiter

The answer is D.
The temperature on the surface of Venus is about 465°C.

103. The mass of the Milky Way is mainly composed by

    A. Dark matter

    B. Stars

    C. Planets

    D. Dust

    E. Asteroids

The answer is A.
It is estimated that 90% of the total mass of the Milky Way is formed by dark matter.

104. The light that reaches the planets originates in the

    A. Dwarf planets

    B. Black holes

    C. Asteroids

    D. Moons

    E. Stars

The answer is E.
Dwarf planets, asteroids, and moons are celestial bodies that don't emit visible light. Black holes are regions with strong gravitational effects. Only stars emit visible light, due to their extremely high temperatures.

NATURAL SCIENCES

105. In 2015, the *New Horizons* spacecraft flew by Pluto, a dwarf planet of the Solar System. How many moons does Pluto have?

   A. Pluto doesn't have moons.

   B. Two moons

   C. Three moons

   D. Four moons

   E. Five moons

The answer is E.
The moons of Pluto are Charon, Styx, Nix, Kerberos, and Hydra.

106. Geostationary satellites are often used by communications. Their orbital period is ____ Earth's rotational period, and their direction of rotation is ____

   A. Shorter than – the same as Earth's

   B. Longer than – opposite to Earth's

   C. Equal to – the same as Earth's

   D. Shorter than – opposite to Earth's

   E. Longer than – the same as Earth's

The answer is C.
Seen by a ground-based observer, geostationary satellites appear to be always in the same position. This is only possible because their orbital period is equal to Earth's rotational period, and their direction of rotation is the same.

NATURAL SCIENCES

107. Which list shows only terrestrial planets?

    A. Earth, Mercury, Jupiter, and Mars

    B. Earth, Mercury, Venus, and Mars

    C. Jupiter, Saturn, Neptune, and Uranus

    D. Earth, Mercury, Neptune, and Uranus

    E. Earth, Mercury, Venus, and Uranus

The answer is B.
These four planets are mainly composed of rock and metals. Jupiter and Saturn are gas giants, composed mainly of hydrogen and helium, and Uranus and Neptune are ice giants.

108. Why Mars is called "the red planet"?

    A. Its waters are rich in sulfur compounds

    B. Its waters are rich in iron compounds

    C. Due to gases present in the atmosphere

    D. Iron oxides in its surface give a reddish appearance

    E. Its surface is rich in sulfur compounds

The answer is D.
Mars' surface is rich in iron oxides, not sulfur. Bodies of water are unknown to occur in Mars.

109. Which layer of the atmosphere contains the ozone layer?

    A. Stratosphere

    B. Thermosphere

    C. Mesosphere

    D. Troposphere

    E. Ionosphere

The answer is A.
Although ozone (O_3) molecules are spread in the atmosphere, they occur in higher concentrations in the stratosphere and originate a layer of this gas.

NATURAL SCIENCES

110. What is the most abundant gas in our atmosphere?

    A. Oxygen

    B. Nitrogen

    C. Helium

    D. Argon

    E. Carbonic dioxide

The answer is B.
Our atmosphere contains 78% nitrogen, 20% oxygen, 1% argon, 0.04% carbon dioxide, and traces of other gases.

111. Which process adds oxygen to the atmosphere?

    A. Respiration

    B. Ultraviolet light

    C. Weathering of rocks

    D. Photosynthesis

    E. Movements of air masses

The answer is D.
Oxygen is a byproduct of photosynthesis. Respiration removes oxygen from the air, and ultraviolet light can break up O_2 into atomic oxygen (O). Air masses move the oxygen but don't create it. The weathering of rocks is not known to produce oxygen.

NATURAL SCIENCES

112. What is the impact of sulfur oxides and nitrogen oxides in the environment when they react with water?

    A. Ammonia

    B. Acidic precipitation

    C. Carbonic acid

    D. Global warming

    E. Greenhouse effect

The answer is B.
When combined with water, sulfur oxides and nitrogen oxides produce acids that lower the pH of the precipitated water, forming what is called "acid rain."

113. Most of Earth's freshwater is

    A. Available for human consumption

    B. Found in rivers and lakes

    C. Found in the groundwater

    D. Found in the clouds and atmosphere

    E. Trapped in the form of ice

The answer is E.
Although rivers and lakes are abundant in many areas of the continents, they account for only 0.3% of total fresh water. About 69% of the Earth's fresh water exist in the form of ice, in the Arctic, Antarctic, and alpine regions.

NATURAL SCIENCES

114. Which of the following statements about groundwater is true?

I. Groundwater is a renewable resource
II. Groundwater cannot be replaced by sea water
III. Groundwater never flows to the surface of the Earth

A. I only

B. II only

C. III only

D. I and II

E. I, II and III

The answer is D.
The groundwater participates in the water cycle, collecting from and exporting water to the surface and thus it is a renewable resource. Groundwater is exclusively fresh water.

115. What are the current Era, Period and Epoch we are living in, respectively?

A. Cenozoic, Quaternary, Holocene

B. Mesozoic, Quaternary, Holocene

C. Mezozoic, Quaternary, Pleistocene

D. Cenozoic, Neogene, Holocene

E. Cenozoic, Neogene, Eocene

The answer is A.
The Mesozoic era took place about 252 to 66 million years ago. The Neogene was previous to the Quaternary. The Eocene was an epoch of the Paleogene, and the Pleistocene was the epoch that preceded the Holocene.

NATURAL SCIENCES

116. What is the main element present in the Earth's core?

    A. Carbon

    B. Sulfur

    C. Iron

    D. Cobalt

    E. Manganese

The answer is C.
The earth's core is mainly composed of iron, and to a lesser extent, of nickel.

117. The lava released by volcanoes during eruption originates in

    A. The Earth's inner mantle

    B. The volcano itself

    C. The Earth's outer core

    D. The Earth's inner core

    E. Magma chambers

The answer is E.
Magma chambers occur in the Earth's crust, commonly under volcanoes.

118. Which statement about the carbon cycle is NOT true?

    A. Ten percent of all available carbon is in the air.

    B. Carbon dioxide is fixed by glycosylation.

    C. Plants fix carbon in the form of glucose.

    D. Animals release carbon through respiration.

    E. Most atmospheric carbon comes from the decay of dead organisms.

The answer is B.
Carbon dioxide is fixed by photosynthesis.

NATURAL SCIENCES

119. The supercontinent Pangea was fragmented due to which geological process?

A. Sedimentation

B. Erosion

C. Chemical weathering

D. Plate tectonics

E. Meteorite impact

The answer is D.
Plate tectonics is the movement of large portions of the Earth's lithosphere.

120. Which statement(s) about fossils is (are) true?

I. Only animals with bones can become fossils
II. There are insect fossils found in amber
III. The fossil's age can be estimated by radiometric aging
IV. The fossil's age can be estimated by stratigraphy

A. I only

B. II only

C. I and II

D. II, III and IV

E. I, II, III and IV

The answer is D.
Many other structures of living organisms can leave a fossil record, for instance, mollusk shells, the exoskeleton of arthropods, teeth, eggs, leaves, and more.

BIOLOGY

Description of the Examination
The Biology examination covers material that is usually taught in a one-year college general biology course. The subject matter tested covers the broad field of the biological sciences, organized into three major areas: molecular and cellular biology, organismal biology, and population biology.

The examination gives approximately equal weight to these three areas. The examination contains approximately 115 questions to be answered in 90 minutes. Some of these are pretest questions that will not be scored. Any time candidates spend on tutorials and providing personal information is in addition to the actual testing time.

Knowledge and Skills Required
Questions on the Biology examination require candidates to demonstrate one or more of the following abilities.

- Knowledge of facts, principles, and processes of biology
- Understanding the means by which information is collected, how it is interpreted, how one hypothesizes from available information, how one draws conclusions and makes further predictions
- Understanding that science is a human endeavor with social consequences

The subject matter of the Biology examination is drawn from the following topics. The percentages next to the main topics indicate the approximate percentage of exam questions on that topic.

33% Molecular and Cellular Biology
- Chemical composition of organisms
- Simple chemical reactions and bonds
- Properties of water
- Chemical structure of carbohydrates, lipids, proteins, nucleic acids
- Origin of life

Cells
- Structure and function of cell organelles
- Properties of cell membranes
- Comparison of prokaryotic and eukaryotic cells

Enzymes
- Enzyme-substrate complex
- Roles of coenzymes
- Inorganic cofactors
- Inhibition and regulation

Energy transformations
- Glycolysis, respiration, anaerobic pathways
- Photosynthesis

Cell division
- Structure of chromosomes
- Mitosis, meiosis, and cytokinesis in plants and animals

Chemical nature of the gene
- Watson-Crick model of nucleic acids
- DNA replication
- Mutations
- Control of protein synthesis: transcription, translation, posttranscriptional processing
- Structural and regulatory genes
- Transformation
- Viruses

34% Organismal Biology
- Structure and function in plants with emphasis on angiosperms
- Root, stem, leaf, flower, seed, fruit
- Water and mineral absorption and transport
- Food translocation and storage
- Plant reproduction and development
- Alternation of generations in ferns, conifers, and flowering plants
- Gamete formation and fertilization
- Growth and development: hormonal control
- Tropisms and photoperiodicity

Structure and function in animals with emphasis on vertebrates
- Major systems (e.g., digestive, gas exchange, skeletal, nervous, circulatory, excretory, immune)
- Homeostatic mechanisms
- Hormonal control in homeostasis and reproduction

Animal reproduction and development
- Gamete formation, fertilization
- Cleavage, gastrulation, germ layer formation, differentiation of organ systems
- Experimental analysis of vertebrate development
- Extraembryonic membranes of vertebrates
- Formation and function of the mammalian placenta
- Blood circulation in the human embryo

Principles of heredity
- Mendelian inheritance (dominance, segregation, independent assortment)
- Chromosomal basis of inheritance
- Linkage, including sex-linked
- Polygenic inheritance (height, skin color)

33% Population Biology
Principles of ecology
- Energy flow and productivity in ecosystems
- Biogeochemical cycles
- Population growth and regulation (natality, mortality, competition, migration, density, r- and K-selection)
- Community structure, growth, regulation (major biomes and succession)
- Habitat (biotic and abiotic factors)
- Concept of niche
- Island biogeography
- Evolutionary ecology (life history strategies, altruism, kin selection)

Principles of evolution
- History of evolutionary concepts
- Concepts of natural selection (differential reproduction, mutation, Hardy-Weinberg equilibrium, speciation, punctuated equilibrium)
- Adaptive radiation
- Major features of plant and animal evolution
- Concepts of homology and analogy
- Convergence, extinction, balanced polymorphism, genetic drift

- Classification of living organisms
- Evolutionary history of humans

Principles of behavior
- Stereotyped, learned social behavior
- Societies (insects, birds, primates)

Social biology
- Human population growth (age composition, birth and fertility rates, theory of demographic transition)
- Human intervention in the natural world (management of resources, environmental pollution)
- Biomedical progress (control of human reproduction, genetic engineering)

BIOLOGY

SAMPLE TEST

DIRECTIONS: Read each item and select the best response.

1. **Which is not true about a cell membrane?**
   (Molecular & Cell Biology)

   A. It is made from phospholipids

   B. Both plant and animal cells have a cell membrane.

   C. The cell wall is the same as the cell membrane in plants.

   D. It controls the passage of nutrients within a cell.

   E. It contains embedded proteins that help with passage.

2. **Microorganisms use all but which of the following for locomotion?**
   (Organismal Biology)

   A. Pseudopods

   B. Flagella

   C. Cilia

   D. Pili

   E. Villi

3. **Which of the following does not possess eukaryotic cells?**
   (Organismal Biology)

   A. Bacteria

   B. Protists

   C. Fungi

   D. Animals

   E. Plants

4. **Which of the following groups of organisms is comprised of those with one cell and no nuclear membrane?**
   (Organismal Biology)

   A. Monera

   B. Protista

   C. Fungi

   D. Algae

   E. Plantae

BIOLOGY

5. **Which of these are found on the outside of the rough endoplasmic reticulum?**
   (Molecular & Cell Biology)

   A. Vacuoles

   B. Mitochondria

   C. Microfilaments

   D. Ribosomes

   E. Flagella

6. **Identify the correct sequence of organization of living things.**
   (Molecular & Cell Biology)

   A. cell – organelle – organ – tissue – organ system – organism

   B. cell – tissue – organ – organelle – organ system – organism

   C. organelle – cell – tissue – organ – organ system – organism

   D. organ system – tissue – organelle – cell – organism – organ

   E. organism – organ system – tissue – cell – organelle – organ

7. **Which of these is not a characteristic shared by all living things?**
   (Organismal Biology)

   A. movement

   B. made of cells

   C. metabolism

   D. reproduction

   E. respond to stimuli

8. **What is the purpose of the Golgi apparatus?**
   (Molecular & Cell Biology)

   A. To break down proteins

   B. To break down fats

   C. To make carbohydrates.

   D. To provide the cell with energy

   E. To sort, modify and package molecules

9. **What do amyloplasts do?** *(Molecular & Cell Biology)*

   A. Store starch in a plant cell

   B. Remove waste in animal cells

   C. Produce green and yellow pigment

   D. Aid in photosynthesis.

   E. Provide energy for metabolism

BIOLOGY

10. **Which of the following does not belong to the domain Archaea?**
    (Organismal Biology)

    A. Methanogens

    B. Extreme Halophiles

    C. Thermoacidophiles

    D. Bacteriophiles

    E. Sulfobales

11. **The first cells that evolved on earth were probably of which type?**
    (Population Biology)

    A. autotrophic

    B. eukaryotic

    C. heterotrophic

    D. prokaryotic

    E. endosymbiotic

12. **During which part of photosynthesis is oxygen given off?**
    (Molecular & Cell Biology)

    A. light reactions

    B. dark reactions

    C. Krebs cycle

    D. reduction of NAD+ to NADH

    E. phosphorylation

13. **Bacteria commonly reproduce by a process called binary fission. Which of the following best defines this process?**
    (Organismal Biology)

    A. Viral vectors carry DNA to new bacteria.

    B. DNA from one bacterium enters another.

    C. DNA doubles and the bacterial cell divides.

    D. DNA from dead cells is absorbed into bacteria.

    E. Bacteria merge with others to form new species.

14. **Which tool is best for studying the individual parts of cells?**
    (Molecular & Cell Biology)

    A. ultracentrifuge

    B. phase-contrast microscope

    C. CAT scan

    D. electron microscope

    E. light microscope

BIOLOGY

15. **Which of these classifications includes the thermoacidophiles?**
    (Organismal Biology)

    A. Plantae

    B. Animalia

    C. Bacteria

    D. Protista

    E. Archaea

16. **Which of the following is not part of the cytoskeleton?**
    (Molecular & Cell Biology)

    A. vacuoles

    B. microfilaments

    C. microtubules

    D. intermediate filaments

    E. motor proteins

17. **Of what are viruses made?**
    (Molecular & Cell Biology)

    A. A protein coat surrounding a nucleic acid.

    B. RNA and protein surrounded by a cell wall.

    C. A nucleic acid surrounding a protein coat.

    D. Protein surrounded by DNA.

    E. A lipid bilayer surrounding a protein coat and RNA.

18. **Which of these are used to classify protists into their major groups?**
    (Organismal Biology)

    A. Their method of obtaining nutrition.

    B. Their method of reproduction.

    C. Their use of metabolism.

    D. Their form and function.

    E. Their means of locomotion.

19. **Replication of chromosomes occurs during which phase of the cell cycle?**
    (Molecular & Cell Biology)

    A. prophase

    B. interphase

    C. metaphase

    D. anaphase

    E. metaphase

20. **Which of these events occurs during telophase in a plant cell?**
    (Molecular & Cell Biology)

    A. the chromosomes are doubled

    B. a cell plate forms

    C. crossing over occurs

    D. a cleavage furrow develops

    E. spindle fibers become visible

BIOLOGY

21. **What is the stage of mitosis seen in the diagram?**
    (Molecular & Cell Biology)

    A. anaphase

    B. metaphase

    C. telophase

    D. prophase

    E. interphase

22. **What is the stage of mitosis shown in the diagram?**
    (Molecular & Cell Biology)

    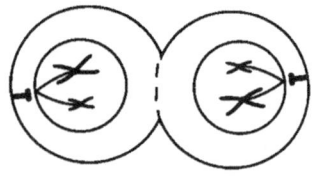

    A. prophase

    B. telophase

    C. anaphase

    D. metaphase

    E. interphase

23. **What is the stage of mitosis shown in the diagram?**
    (Molecular & Cell Biology)

    A. interphase

    B. metaphase

    C. prophase

    D. telophase

    E. anaphase

24. **Which of the following is a monomer?**
    (Molecular & Cell Biology)

    A. RNA

    B. glycogen

    C. DNA

    D. amino acid

    E. lipid

25. **Which of the following does not affect enzyme rate?**
    (Molecular & Cell Biology)

    A. increase of temperature

    B. amount of substrate

    C. pH

    D. size of the cell

    E. concentration of enzyme

BIOLOGY

26. **All but which one of the following is true of a cell membrane?**
    (Molecular & Cell Biology)

    A. It contains polar and nonpolar phospholipids.

    B. It only uses active transport to move molecules across it.

    C. It contains cholesterol.

    D. It has proteins imbedded within it.

    E. It is selectively permeable to many substances.

27. **Which of these describes facilitated diffusion?**
    (Molecular & Cell Biology)

    A. It requires energy.

    B. It only happens in plant cells.

    C. It only allows molecules to leave a cell but not to enter it.

    D. It produces a significant amount of energy for the cell.

    E. It needs a transport molecule to pass through the membrane.

28. **What is not true of enzymes?**
    (Molecular & Cell Biology)

    A. They are the most diverse of all proteins.

    B. They act on a substrate.

    C. They work at a wide range of pH.

    D. They are temperature-dependent.

    E. They have specialized functions.

29. **Which of these is necessary for diffusion to occur?**
    (Molecular & Cell Biology)

    A. carrier proteins

    B. energy

    C. water molecules

    D. a cell membrane

    E. a concentration gradient

30. **Which is an example of the use of energy to move a substance through a membrane from areas of low concentration to areas of high concentration?**
    (Molecular & Cell Biology)

    A. osmosis

    B. active transport

    C. exocytosis

    D. phagocytosis

    E. facilitated diffusion

BIOLOGY

31. **A plant cell is placed in salt water. What is the resulting movement of water out of the cell called?**
    (Molecular & Cell Biology)

    A. facilitated diffusion

    B. diffusion

    C. transpiration

    D. osmosis

    E. active transport

32. **What are the monomers of polysaccharides?**
    (Molecular & Cell Biology)

    A. Nucleotides

    B. Amino acids

    C. Polypeptides

    D. Fatty acids

    E. Simple sugars

33. **Which type of cell would contain the most mitochondria?**
    (Molecular & Cell Biology)

    A. muscle cell

    B. nerve cell

    C. epithelial cell

    D. blood cell

    E. bone cell

34. **According to the fluid-mosaic model of the cell membrane, of what are membranes composed?**
    (Molecular & Cell Biology)

    A. Phospholipid bilayers with proteins embedded in the layers.

    B. One layer of phospholipids with cholesterol embedded in the layer.

    C. Two layers of protein with lipids embedded in the layers.

    D. DNA and fluid proteins with carbohydrates embedded in the layer.

    E. Glycerol and RNA with carbohydrates embedded in the layer.

BIOLOGY

35. **Which is the correct statement regarding the human nervous system and the human endocrine system?**
    (Organismal Biology)

    A. The nervous system maintains homeostasis whereas the endocrine system does not.

    B. Endocrine glands produce neurotransmitters whereas nerves produce hormones.

    C. Nerve signals travel on neurons whereas hormones travel through the blood.

    D. The nervous system involves chemical transmission whereas the endocrine system does not.

    E. The nervous system produces physiological responses whereas the endocrine produces behavioral.

36. **Which process generates the most ATP?**
    (Molecular & Cell Biology)

    A. fermentation

    B. glycolysis

    C. the Calvin cycle

    D. the Krebs cycle

    E. chemiosmosis

37. **Which of these is a function of the cardiovascular system?**
    (Organismal Biology)

    A. Move oxygenated blood around the body

    B. Oxygenate the blood through gas exchange

    C. Act as an exocrine system

    D. Flush toxins out of the body

    E. Transport signals from the brain

38. **Which of these is not a part of the nervous system?**
    (Organismal Biology)

    A. brain

    B. spinal cord

    C. axons

    D. venules

    E. cochlea

39. **Organisms need to maintain a constant internal environment to survive. Which of these is a method by which they achieve this?**
    (Organismal Biology)

    A. respiration

    B. reproduction

    C. depolarization

    D. repolarization

    E. thermoregulation

BIOLOGY

40. Which of these controls the body's endocrine mechanisms?
    (Organismal Biology)

    A. feedback loops

    B. control molecules

    C. neurochemicals

    D. neurotransmitters

    E. behavioral responses

41. What is the gland that regulates the calcium in the body?
    (Organismal Biology)

    A. Thyroid gland

    B. Parathyroid gland

    C. Hypothalamus

    D. Pituitary gland

    E. Pancreas

42. Which of these steroids is not created in the gonads?
    (Organismal Biology)

    A. Testosterone

    B. Estrogen

    C. Progesterone

    D. ACTH

    E. FSH

43. What is the most common neurotransmitter?
    (Organismal Biology)

    A. epinephrine

    B. serotonin

    C. acetyl choline

    D. norepinephrine

    E. oxytocin

44. Food is carried through the digestive tract by a series of wave-like contractions. What is this process is called?
    (Organismal Biology)

    A. peristalsis

    B. chyme

    C. digestion

    D. absorption

    E. depolarization

45. Which of these must muscles pull on in order to initiate movement?
    (Organismal Biology)

    A. skin

    B. bones

    C. joints

    D. ligaments

    E. bursa

BIOLOGY

46. Hormones are essential to the regulation of reproduction. What organ is responsible for the release of hormones for sexual maturity?
    (Organismal Biology)

    A. pituitary gland

    B. hypothalamus

    C. pancreas

    D. thyroid gland

    E. pineal gland

47. What is the type of muscle in the human body that is voluntary?
    (Organismal Biology)

    A. Cardiac

    B. Sarcomere

    C. Smooth

    D. Skeletal

    E. Actin

48. The wrist is an example of what kind of joint?
    (Organismal Biology)

    A. Ball and socket

    B. Pivot

    C. Stationary

    D. Hinge

    E. Gliding

49. What is the waterproofing protein in the skin called?
    (Organismal Biology)

    A. actin

    B. epidermis

    C. collagen

    D. sebum

    E. keratin

50. What is the muscular adaptation called that is used to move food through the digestive system?
    (Organismal Biology)

    A. peristalsis

    B. passive transport

    C. voluntary action

    D. bulk transport

    E. endocytosis

51. What is the role of neurotransmitters in nerve action?
    (Organismal Biology)

    A. to turn off the sodium pump

    B. to turn off the calcium pump

    C. to send impulses to neurons

    D. to send impulses around the body

    E. to send impulses from axon to dendrite

BIOLOGY

52. **Fats are broken down by which substance?**
    (Organismal Biology)

    A. bile produced in the gall bladder

    B. lipase produced in the gall bladder

    C. glucagons produced in the liver

    D. amylase produces in the gall bladder

    E. bile produced in the liver

53. **Where does fertilization in humans usually occurs?**
    (Organismal Biology)

    A. uterus

    B. ovary

    C. fallopian tubes

    D. vagina

    E. epididymis

54. **Which of these is lacking in the dermis layer of skin?**
    (Organismal Biology)

    A. sweat glands

    B. keratin

    C. hair follicles

    D. blood vessels

    E. living cells

55. **A school age boy had the chicken pox as a baby. Why will he most likely not get this disease again?**
    (Organismal Biology)

    A. passive immunity

    B. vaccination

    C. antibiotics

    D. active immunity

    E. antigen production

56. **What is any foreign particle called that causes an immune reaction?**
    (Organismal Biology)

    A. an antigen

    B. a histocompatibity complex

    C. an antibody

    D. a vaccine

    E. a bacteriophage

57. Which of these statements describes the polymerase chain reaction?
    (Population Biology)

    A. It is a group of polymerases.

    B. It is a technique for amplifying DNA.

    C. It is a primer for DNA synthesis.

    D. It is a way to synthesize polymerase.

    E. It is a series of genetic mutations.

58. Which part of a DNA nucleotide can vary?
    (Molecular & Cell Biology)

    A. deoxyribose

    B. phosphate group

    C. hydrogen bonds

    D. sugar

    E. nitrogenous base

59. A DNA strand has the base sequence of TCAGTA. Its DNA complement would have which of the following sequences?
    (Molecular & Cell Biology)

    A. ATGACT

    B. TCAGTA

    C. AGUCAU

    D. AGTCAT

    E. TCTGTA

60. Which of these carries amino acids to the ribosome during protein synthesis?
    (Molecular & Cell Biology)

    A. messenger RNA

    B. ribosomal RNA

    C. transfer RNA

    D. DNA

    E. RNA

61. A protein is sixty amino acids in length. This requires a coded DNA sequence of how many nucleotides?
    (Molecular & Cell Biology)

    A. 20

    B. 30

    C. 120

    D. 180

    E. 240

62. A DNA molecule has the sequence of ACTATG. What is the anticodon of this molecule?
    (Molecular & Cell Biology)

    A. UGAUAC

    B. ACUAUG

    C. TGATAC

    D. ACTATG

    E. CTGCGA

BIOLOGY

63. What is the general term for a change that affects the sequence of bases in a gene?
    (Molecular & Cell Biology)

    A. deletion

    B. polyploid

    C. mutation

    D. duplication

    E. substitution

64. Segments of DNA can be transferred from the DNA of one organism to another through the use of which of the following?
    (Population Biology)

    A. bacterial plasmids

    B. viruses

    C. chromosomes from frogs

    D. plant DNA

    E. Okazaki fragments

65. What is the enzyme that unwinds DNA during replication?
    (Molecular & Cell Biology)

    A. DNAse

    B. DNA replicase

    C. DNA helicase

    D. DNA topoisomerases

    E. DNA polymerase

66. What is a small circular piece of DNA called that contains accessory DNA?
    (Molecular & Cell Biology)

    A. mitochondrial DNA

    B. messenger RNA

    C. transfer DNA

    D. Okazaki fragment

    E. plasmid

67. In DNA, adenine bonds with _____, while cytosine bonds with _____.
    (Molecular & Cell Biology)

    A. thymine/guanine

    B. adenine/cytosine

    C. cytosine/uracil

    D. guanine/thymine

    E. uracil/adenine

68. Which protein structure consists of the coils and folds of polypeptide chains?
    (Molecular & Cell Biology)

    A. secondary structure

    B. quaternary structure

    C. tertiary structure

    D. primary structure

    E. quinary structure

BIOLOGY

69. **What can be said about homozygous individuals?**
    (Organismal Biology)

    A. They have two different alleles.

    B. They are of the same species.

    C. They exhibit the same features.

    D. They have a pair of identical alleles.

    E. They produce identical offspring.

70. **The term "phenotype" refers to which of the following?**
    (Organismal Biology)

    A. a condition that is heterozygous

    B. the genetic makeup of an individual

    C. a condition that is homozygous

    D. how the genotype is expressed

    E. from which parent the traits were inherited

71. **The ratio of brown-eyed to blue-eyed children from the mating of a blue-eyed male to a heterozygous brown-eyed female is expected to be which of the following?**
    (Organismal Biology)

    A. 3:1

    B. 2:2

    C. 1:0

    D. 1:2

    E. 0:4

72. **Which of these defines the Law of Segregation defined by Gregor Mendel?**
    (Organismal Biology)

    A. After meiosis, each new cell will contain an allele that is recessive.

    B. Only one of two alleles is expressed in a heterozygous organism.

    C. The allele expressed is always the dominant allele.

    D. Alleles of one trait do not affect the inheritance of alleles on another chromosome.

    E. When sex cells form, the two alleles that determine a trait will end up on different gametes.

BIOLOGY

73. Which of the following is an example of the incomplete dominance that occurs when a white flower is crossed with a red flower?
    (Organismal Biology)

    A. pink flowers

    B. red flowers

    C. white flowers

    D. red and white flowers

    E. white and pink flowers

74. A child with type O blood has a father with type A blood and a mother with type B blood. The genotypes of the parents respectively would be which of the following?
    (Organismal Biology)

    A. AA and BO

    B. AO and BO

    C. AA and BB

    D. AO and OO

    E. OO and AB

75. Crossing over, which increases genetic diversity, occurs during which stage(s) of meiosis?
    (Molecular & Cell Biology)

    A. telophase II in meiosis

    B. metaphase in mitosis

    C. interphase in meiosis

    D. prophase I in meiosis

    E. metaphase II in meiosis

76. ABO blood grouping is an example of which type of allele dominance?
    (Organismal Biology)

    A. Autosomal dominance

    B. Incomplete dominance

    C. Somatic dominance

    D. Complete dominance

    E. Codominance

77. In a Punnett square with a single trait, what are the ratios of genotypes produced between two heterozygous individuals?
    (Organismal Biology)

    A. 1:2:2

    B. 2:1:1

    C. 1:1:1

    D. 1:2:1

    E. 2:2:2

BIOLOGY

78. **What is the term for an organism's genetic makeup?**
    (Organismal Biology)

    A. Heterozygote

    B. Genotype

    C. Phenotype

    D. Homozygote

    E. Dominance

79. **Which of these represents a genetic engineering advancement in the medical field?**
    (Population Biology)

    A. stem cell reproduction

    B. pesticides

    C. degradation of harmful chemicals

    D. antibiotics

    E. gene therapy

80. **Which of the following is not true regarding restriction enzymes?**
    (Population Biology)

    A. They aid in transcombination procedures.

    B. They are used in genetic engineering.

    C. They are named after the bacteria in which they naturally occur.

    D. They identify and splice certain base sequences on DNA.

    E. They can be produced by certain lipids during DNA replication.

81. **Which of these processes is not one of the modern uses of DNA?**
    (Population Biology)

    A. PCR technology

    B. Gene therapy

    C. Cloning

    D. Genetic Alignment

    E. Transgenic organisms

BIOLOGY

82. **Which statement best represents gel electrophoresis?**
    (Population Biology)

    A. It isolates fragments of DNA for scientific purposes.

    B. It cannot be used in proteins.

    C. It requires the polymerase chain reaction.

    D. It only separates DNA by size.

    E. It uses different charged particles to color the bands.

83. **What is the term that describes the duplication of genetic material into another cell?**
    (Population Biology)

    A. replicating

    B. cell duplication

    C. transgenics

    D. genetic restructuring

    E. cloning

84. **What does gel electrophoresis use to separate the DNA?**
    (Population Biology)

    A. the amount of current

    B. the size of the molecule

    C. the positive charge of the molecule

    D. the solubility of the gel

    E. the source of the DNA

85. **Which of these is a result of reproductive isolation?**
    (Population Biology)

    A. extinction

    B. migration

    C. fossilization

    D. speciation

    E. radiation

86. **Which of these is true about natural selection?**
    (Population Biology)

    A. It acts on an individual genotype.

    B. It is not currently happening.

    C. It is only an animal phenomenon.

    D. It acts on the individual phenotype.

    E. It is used to prevent overpopulation.

87. **How does diversity aid a population?**
    (Population Biology)

    A. Individuals are better able to survive.

    B. Mates are attracted to a diverse population.

    C. Potential mates like conformity.

    D. It increases the DNA differences in the population.

    E. It provides possible improvements to the population.

BIOLOGY

88. **Which statement is not true about diversity?**
    (Population Biology)

    A. Without diversity there would be extinction.

    B. Diversity is increasing all the time.

    C. Fossil evidence supports diversity.

    D. Sexual reproduction encourages more diversity.

    E. Skeletons are too similar to allow for diversity.

89. **Which of these ideas was a major part of Darwin's evolutionary theory?**
    (Population Biology)

    A. Punctualism

    B. Gradualism

    C. Equilibrium

    D. Convergency

    E. Altruism

90. **Which statement is not true about reproductive isolation?**
    (Population Biology)

    A. It prevents populations from exchanging genes.

    B. It can occur by preventing fertilization.

    C. It can result in speciation.

    D. It happens more often on the mainland.

    E. It produces offspring with unique phenotypes

91. **Which idea is true about members of the same species?**
    (Population Biology)

    A. They look identical.

    B. They never change.

    C. They reproduce successfully within their group.

    D. They live in the same geographic location.

    E. They have very dissimilar genotypes.

BIOLOGY

92. Which of the following factors will affect the Hardy-Weinberg law of equilibrium, leading to evolutionary change?
(Population Biology)

    A. no mutations

    B. non-random mating

    C. no immigration or emigration

    D. large population

    E. small individual species

93. If a population is in Hardy-Weinberg equilibrium and the frequency of the recessive allele is 0.3, what percentage of the population is expected to be heterozygous?
(Population Biology)

    A. 9%

    B. 49%

    C. 42%

    D. 21%

    E. 7%

94. Which aspect of science does not support evolution?
(Population Biology)

    A. comparative anatomy

    B. organic chemistry

    C. comparison of DNA among organisms

    D. analogous structures

    E. embryology

95. In which of these does evolution occurs?
(Population Biology)

    A. individuals

    B. populations

    C. organ systems

    D. cells

    E. ecosystems

96. Which process contributes most to the large variety of living things in the world today?
(Population Biology)

    A. meiosis

    B. asexual reproduction

    C. mitosis

    D. alternation of generations

    E. reproductive isolation

BIOLOGY

97. Which of the following gases was a major part of the primitive Earth atmosphere?
    (Population Biology)

    A. fluorine

    B. methane

    C. oxygen

    D. krypton

    E. argon

98. What is a major principle of the Endosymbiotic Theory?
    (Population Biology)

    A. Birds and dinosaurs share a common ancestor.

    B. Animals evolved in close relationships with one another.

    C. Prokaryotes arose from eukaryotes.

    D. Inorganic compounds are the basis of living things.

    E. Eukaryotes arose from very simple prokaryotes.

99. The wing of a bird, the human arm, and the pectoral fluke of a whale all have the same bone structure. What are these structures called?
    (Population Biology)

    A. polymorphic structures

    B. homologous structures

    C. vestigial structures

    D. analogous structures

    E. allopatric structures

100. Which of the following is not an abiotic factor?
    (Population Biology)

    A. temperature

    B. rainfall

    C. soil quality

    D. predation

    E. wind speed

BIOLOGY

101. What is not true about cladistics?
(Population Biology)

A. It is the study of phylogenetic relationships of organisms.

B. It involves a branching diagram that uses the development of novel traits to separate groups of organisms.

C. It distinguishes between the relative importance of the traits.

D. It shows when traits developed with respect to other traits.

E. It indicates which organisms are most closely related to each other and what their common ancestors were.

102. If DDT were present in an ecosystem, which of the following organisms would have the highest concentration in its body?
(Population Biology)

A. herring

B. diatom

C. zooplankton

D. salmon

E. osprey

103. What eats secondary consumers?
(Population Biology)

A. Producers

B. Tertiary consumers

C. Primary consumers

D. Decomposers

E. Detritivores

104. Which statement is true about the water cycle?
(Population Biology)

A. Two percent of the water is fixed and unavailable.

B. 75% of available water is groundwater.

C. The water cycle is driven by the ocean currents.

D. Surface water percolates up from underground springs.

E. New water is being added into the cycle all the time.

BIOLOGY

105. Which statement about the carbon cycle is false?
(Population Biology)

A. Ten percent of all available carbon is in the air.

B. Carbon dioxide is fixed by glycosylation.

C. Plants fix carbon in the form of glucose.

D. Animals release carbon through respiration.

E. Most atmospheric carbon comes from the decay of dead organisms.

106. What is the impact of sulfur oxides and nitrogen oxides in the environment when they react with water?
(Population Biology)

A. ammonia

B. acidic precipitation

C. sulfuric acid

D. global warming

E. greenhouse effect

107. Which term is not associated with the water cycle?
(Population Biology)

A. precipitation

B. transpiration

C. fixation

D. evaporation

E. runoff

108. Which of the following is a density dependent factor that affects a population?
(Population Biology)

A. temperature

B. rainfall

C. predation

D. soil nutrients

E. wind speed

109. High humidity and temperature stability are present in which of the following biomes?
(Population Biology)

A. taiga

B. deciduous forest

C. desert

D. tropical rain forest

E. coniferous forest

BIOLOGY

110. Which trophic level has the highest ecological efficiency?
(Population Biology)

A. decomposers

B. producers

C. tertiary consumers

D. secondary consumers

E. primary consumers

111. From where does the oxygen created in photosynthesis come?
(Molecular & Cell Biology)

A. carbon dioxide

B. chlorophyll

C. glucose

D. carbon monoxide

E. water

112. Which of the following is true of decomposers?
(Population Biology)

A. Decomposers recycle the carbon accumulated in durable organic material.

B. They take nitrogen out of the soil to use for food.

C. Decomposers absorb nutrients from the air to maintain their metabolisms.

D. Decomposers belong to the Genus *Escherichia*.

E. They are able to use the Sun to produce their own energy.

113. A clownfish is protected by a sea anemone's tentacles, and in turn, the anemone receives uneaten food from the clownfish. What type of symbiosis is exemplified by this example?
(Population Biology)

A. mutualism

B. parasitism

C. commensalism

D. competition

E. amensalism

BIOLOGY

114. **Which of these is most likely to happen in order for primary succession to occur?**
(Population Biology)

   A. nutrient enrichment

   B. a forest fire

   C. bare rock is exposed after a water table recedes

   D. a housing development is built

   E. a farmer stops cultivating her fields

115. **What is the Mendelian law called that states that only one of the two possible alleles from each parent is passed on to the offspring?**
(Organismal Biology)

   A. The Mendelian Law

   B. The Law of Independent Assortment

   C. The Law of Segregation

   D. The Allele Law

   E. The Law of Dominance and Recessiveness

BIOLOGY

ANSWER KEY

Question Number	Correct Answer	Your Answer	Question Number	Correct Answer	Your Answer	Question Number	Correct Answer	Your Answer
1	C		41	B		81	D	
2	E		42	D		82	A	
3	A		43	C		83	E	
4	A		44	A		84	B	
5	D		45	B		85	D	
6	C		46	B		86	D	
7	A		47	D		87	E	
8	E		48	B		88	E	
9	A		49	E		89	B	
10	D		50	A		90	D	
11	D		51	A		91	C	
12	A		52	E		92	B	
13	C		53	C		93	C	
14	D		54	B		94	B	
15	E		55	D		95	B	
16	A		56	A		96	A	
17	A		57	B		97	B	
18	D		58	E		98	E	
19	B		59	D		99	B	
20	B		60	C		100	D	
21	B		61	D		101	C	
22	B		62	D		102	E	
23	E		63	C		103	B	
24	D		64	A		104	A	
25	D		65	C		105	B	
26	B		66	E		106	B	
27	E		67	A		107	C	
28	C		68	A		108	C	
29	E		69	D		109	D	
30	B		70	D		110	B	
31	D		71	B		111	E	
32	E		72	E		112	A	
33	A		73	A		113	A	
34	A		74	B		114	C	
35	C		75	D		115	B	
36	E		76	E				
37	A		77	D				
38	D		78	B				
39	E		79	E				
40	A		80	A				

BIOLOGY

RATIONALES

1. **Which is not true about a cell membrane?**
   (Molecular & Cell Biology)

   A. It is made from phospholipids

   B. Both plant and animal cells have a cell membrane.

   C. The cell wall is the same as the cell membrane in plants.

   D. It controls the passage of nutrients within a cell.

   E. It contains embedded proteins that help with passage.

The answer is C.
Both plants and animals have cell membranes but plant cells also have an outer cell wall to give it structure.

2. **Microorganisms use all but which of the following for locomotion?**
   (Organismal Biology)

   A. Pseudopods

   B. Flagella

   C. Cilia

   D. Pili

   E. Villi

The answer is E.
Pseudopods, pili, flagella and cilia are used by microorganisms for movement. Vili are used in the small intestine to increase surface area for absorption.

BIOLOGY

3. **Which of the following does not possess eukaryotic cells?**
    (Organismal Biology)

    A. Bacteria

    B. Protists

    C. Fungi

    D. Animals

    E. Plants

The answer is A.
Eukaryotic cells are found in protists, fungi, plants and animals but not in bacteria.

4. **Which of the following groups of organisms is comprised of those with one cell and no nuclear membrane?**
    (Organismal Biology)

    A. Monera

    B. Protista

    C. Fungi

    D. Algae

    E. Plantae

The answer is A.
Monera is the only kingdom that is made up of unicellular organisms with no nucleus. Algae are protists because it is made up of one type of tissue and it has a nucleus.

BIOLOGY

5. Which of these are found on the outside of the rough endoplasmic reticulum?
(Molecular & Cell Biology)

   A. Vacuoles

   B. Mitochondria

   C. Microfilaments

   D. Ribosomes

   E. Flagella

The answer is D.
Rough endoplasmic reticulum is defined as such because of the occurrence of ribosomes on its surface.

6. Identify the correct sequence of organization of living things.
(Molecular & Cell Biology)

   A. cell – organelle – organ – tissue – organ system – organism

   B. cell – tissue – organ – organelle – organ system – organism

   C. organelle – cell – tissue – organ – organ system – organism

   D. organ system – tissue – organelle – cell – organism – organ

   E. organism – organ system – tissue – cell – organelle – organ

The answer is C.
An organism, such as a human, is comprised of several organ systems such as the circulatory and nervous systems. These organ systems consist of many organs including the heart and the brain. These organs are made of tissue such as cardiac muscle. Tissues are made up of cells, which contain organelles like the mitochondria and the Golgi apparatus.

BIOLOGY

7. **Which of these is not a characteristic shared by all living things?**
(Organismal Biology)

    A. movement

    B. made of cells

    C. metabolism

    D. reproduction

    E. respond to stimuli

The answer is A.
Movement is not a characteristic of life. Viruses are considered non-living organisms but have the ability to move from cell to cell in its host organism. A leaf on a tree or the tree itself are very much alive but unable to move in terms of mobility.

8. **What is the purpose of the Golgi apparatus?**
(Molecular & Cell Biology)

    A. To break down proteins

    B. To break down fats

    C. To make carbohydrates.

    D. To provide the cell with energy

    E. To sort, modify and package molecules

The answer is E.
The Golgi apparatus takes molecules from the endoplasmic reticulum and sorts, modifies and packages the molecules for later use by the cell.

BIOLOGY

9. **What do amyloplasts do?**
   (Molecular & Cell Biology)

   A. Store starch in a plant cell

   B. Remove waste in animal cells

   C. Produce green and yellow pigment

   D. Aid in photosynthesis.

   E. Provide energy for metabolism

The answer is A.
Amyloplasts store starch in plant cells

10. **Which of the following does not belong to the domain Archaea?**
    (Organismal Biology)

    A. Methanogens

    B. Extreme Halophiles

    C. Thermoacidophiles

    D. Bacteriophiles

    E. Sulfobales

The answer is D.
The Archaea group includes all of the above except Bacteriophiles.

BIOLOGY

11. **The first cells that evolved on earth were probably of which type?**
    (Population Biology)

    A. autotrophic

    B. eukaryotic

    C. heterotrophic

    D. prokaryotic

    E. endosymbiotic

The answer is D.
Prokaryotes date back to 3.5 billion years ago in the first fossil record. Their ability to adapt to the environment allows them to thrive in a wide variety of habitats.

12. **During which part of photosynthesis is oxygen given off?**
    (Molecular & Cell Biology)

    A. light reactions

    B. dark reactions

    C. Krebs cycle

    D. reduction of NAD+ to NADH

    E. phosphorylation

The answer is A.
The conversion of solar energy to chemical energy occurs in the light reactions. Electrons are transferred by the absorption of light by chlorophyll and cause water to split, releasing oxygen as a waste product.

BIOLOGY

13. **Bacteria commonly reproduce by a process called binary fission. Which of the following best defines this process?**
    (Organismal Biology)

    A. Viral vectors carry DNA to new bacteria.

    B. DNA from one bacterium enters another.

    C. DNA doubles and the bacterial cell divides.

    D. DNA from dead cells is absorbed into bacteria.

    E. Bacteria merge with others to form new species.

The answer is C.
Binary fission is the asexual process in which the bacteria divide in half after the DNA doubles. This results in an exact clone of the parent cell.

14. **Which tool is best for studying the individual parts of cells?**
    (Molecular & Cell Biology)

    A. ultracentrifuge

    B. phase-contrast microscope

    C. CAT scan

    D. electron microscope

    E. light microscope

The answer is D.
The scanning electron microscope uses a beam of electrons to pass through the specimen. The resolution is about 1000 times greater than that of a light microscope. This allows the scientist to view extremely small objects, such as the individual parts of a cell.

BIOLOGY

15. Which of these classifications includes the thermoacidophiles?
(Organismal Biology)

    A. Plantae

    B. Animalia

    C. Bacteria

    D. Protista

    E. Archaea

The answer is E.
Thermoacidophiles, methanogens, and halobacteria are members of the Archaea group.

16. Which of the following is not part of the cytoskeleton?
(Molecular & Cell Biology)

    A. vacuoles

    B. microfilaments

    C. microtubules

    D. intermediate filaments

    E. motor proteins

The answer is A.
Vacuoles are mostly found in plants and hold stored food and pigments. The other three choices are fibers that make up the cytoskeleton found in both plant and animal cells.

BIOLOGY

17. **Of what are viruses made?**
    (Molecular & Cell Biology)

    A. A protein coat surrounding a nucleic acid.

    B. RNA and protein surrounded by a cell wall.

    C. A nucleic acid surrounding a protein coat.

    D. Protein surrounded by DNA.

    E. A lipid bilayer surrounding a protein coat and RNA.

The answer is A.
Viruses are composed of a protein coat surrounding a nucleic acid; either RNA or DNA.

18. **Which of these are used to classify protists into their major groups?**
    (Organismal Biology)

    A. Their method of obtaining nutrition.

    B. Their method of reproduction.

    C. Their use of metabolism.

    D. Their form and function.

    E. Their means of locomotion.

The answer is D.
The chaotic status of names and concepts of the higher classification of the protists reflects their great diversity in form, function, and life styles. The protists are often grouped as algae (plant-like), protozoa (animal-like), or fungus-like based on the similarity of their lifestyle and characteristics to these more defined groups.

BIOLOGY

19. Replication of chromosomes occurs during which phase of the cell cycle?
(Molecular & Cell Biology)

A. prophase

B. interphase

C. metaphase

D. anaphase

E. metaphase

The answer is B.
Interphase is the stage where the cell grows and copies the chromosomes in preparation for the mitotic phase.

20. Which of these events occurs during telophase in a plant cell?
(Molecular & Cell Biology)

A. the chromosomes are doubled

B. a cell plate forms

C. crossing over occurs

D. a cleavage furrow develops

E. spindle fibers become visible

The answer is B.
During plant cell telophase, a cell plate is observed whereas a cleavage furrow is formed in animal cells.

BIOLOGY

21. **What is the stage of mitosis seen in the diagram?**
    (Molecular & Cell Biology)

    A. anaphase

    B. metaphase

    C. telophase

    D. prophase

    E. interphase

The answer is B.
During metaphase, the centromeres are at opposite ends of the cell. Here the chromosomes are aligned with one another.

22. **What is the stage of mitosis shown in the diagram?**
    (Molecular & Cell Biology)

    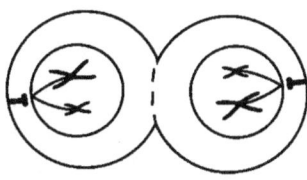

    A. prophase

    B. telophase

    C. anaphase

    D. metaphase

    E. interphase

The answer is B.
Telophase is the last stage of mitosis. Here, two nuclei become visible and the nuclear membrane reassembles.

BIOLOGY

23. What is the stage of mitosis shown in the diagram?
(Molecular & Cell Biology)

    A. interphase

    B. metaphase

    C. prophase

    D. telophase

    E. anaphase

The answer is E.
During anaphase, the centromeres split in half and homologous chromosomes separate.

24. Which of the following is a monomer?
(Molecular & Cell Biology)

    A. RNA

    B. glycogen

    C. DNA

    D. amino acid

    E. lipid

The answer is D.
A monomer is the simplest unit of structure for a particular macromolecule. Amino acids are the basic units that comprise a protein. RNA and DNA are polymers consisting of nucleotides and glycogen is a polymer consisting of many molecules of glucose.

BIOLOGY

25. **Which of the following does not affect enzyme rate?**
    (Molecular & Cell Biology)

    A. increase of temperature

    B. amount of substrate

    C. pH

    D. size of the cell

    E. concentration of enzyme

The answer is D.
Temperature and pH can affect the rate of reaction of an enzyme. The amount of substrate affects the enzyme as well. The enzyme acts on the substrate. The more substrate, the slower the enzyme rate. Therefore, the only choice left is D, the size of the cell, which has no effect on enzyme rate.

26. **All but which one of the following is true of a cell membrane?**
    (Molecular & Cell Biology)

    A. It contains polar and nonpolar phospholipids.

    B. It only uses active transport to move molecules across it.

    C. It contains cholesterol.

    D. It has proteins imbedded within it.

    E. It is selectively permeable to many substances.

The answer is B.
Cell membranes use passive and active transport to transport molecules across the membrane.

BIOLOGY

27. Which of these describes facilitated diffusion?
(Molecular & Cell Biology)

    A. It requires energy.

    B. It only happens in plant cells.

    C. It only allows molecules to leave a cell but not to enter it.

    D. It produces a significant amount of energy for the cell.

    E. It needs a transport molecule to pass through the membrane.

The answer is E.
Facilitated diffusion requires no energy but needs a transport molecule to pass another molecule through the membrane.

28. What is not true of enzymes?
(Molecular & Cell Biology)

    A. They are the most diverse of all proteins.

    B. They act on a substrate.

    C. They work at a wide range of pH.

    D. They are temperature-dependent.

    E. They have specialized functions.

The answer is C.
Enzymes generally work best within a very narrow range in pH.

BIOLOGY

29. **Which of these is necessary for diffusion to occur?**
    (Molecular & Cell Biology)

    A. carrier proteins

    B. energy

    C. water molecules

    D. a cell membrane

    E. a concentration gradient

The answer is E.
Diffusion is the ability of molecules to move from areas of high concentration to areas of low concentration (a concentration gradient).

30. **Which is an example of the use of energy to move a substance through a membrane from areas of low concentration to areas of high concentration?**
    (Molecular & Cell Biology)

    A. osmosis

    B. active transport

    C. exocytosis

    D. phagocytosis

    E. facilitated diffusion

The answer is B.
Active transport can move substances with or against the concentration gradient. This energy-requiring process allows for molecules to move from areas of low concentration to areas of high concentration.

BIOLOGY

31. A plant cell is placed in salt water. What is the resulting movement of water out of the cell called?
(Molecular & Cell Biology)

    A. facilitated diffusion

    B. diffusion

    C. transpiration

    D. osmosis

    E. active transport

The answer is D.
Osmosis is simply the diffusion of water across a semi-permeable membrane. Water will diffuse out of the cell if there is a lower concentration of water on the outside of the cell.

32. What are the monomers of polysaccharides?
(Molecular & Cell Biology)

    A. Nucleotides

    B. Amino acids

    C. Polypeptides

    D. Fatty acids

    E. Simple sugars

The answer is E.
The monomers of polysaccharides are simple sugars.

BIOLOGY

33. Which type of cell would contain the most mitochondria?
(Molecular & Cell Biology)

    A. muscle cell

    B. nerve cell

    C. epithelial cell

    D. blood cell

    E. bone cell

The answer is A.
Mitochondria are the site of cellular respiration where ATP is made. Muscle cells have the most mitochondria because they use a great deal of energy.

34. According to the fluid-mosaic model of the cell membrane, of what are membranes composed?
(Molecular & Cell Biology)

    A. Phospholipid bilayers with proteins embedded in the layers.

    B. One layer of phospholipids with cholesterol embedded in the layer.

    C. Two layers of protein with lipids embedded in the layers.

    D. DNA and fluid proteins with carbohydrates embedded in the layer.

    E. Glycerol and RNA with carbohydrates embedded in the layer.

The answer is A.
Cell membranes are composed of two phospholipids with their hydrophobic tails sandwiched between their hydrophilic heads, creating a lipid bilayer. The membrane contains proteins embedded in the layer (integral proteins) and proteins on the surface (peripheral proteins).

BIOLOGY

35. Which is the correct statement regarding the human nervous system and the human endocrine system?
(Organismal Biology)

A. The nervous system maintains homeostasis whereas the endocrine system does not.

B. Endocrine glands produce neurotransmitters whereas nerves produce hormones.

C. Nerve signals travel on neurons whereas hormones travel through the blood.

D. The nervous system involves chemical transmission whereas the endocrine system does not.

E. The nervous system produces physiological responses whereas the endocrine produces behavioral.

The answer is C.
In the human nervous system, neurons carry nerve signals to and from the cell body. Endocrine glands produce hormones that are carried through the body in the bloodstream.

36. Which process generates the most ATP?
(Molecular & Cell Biology)

A. fermentation

B. glycolysis

C. the Calvin cycle

D. the Krebs cycle

E. chemiosmosis

The answer is E.
The electron transport chain uses electrons to pump hydrogen ions across the mitochondrial membrane. This ion gradient is used to form ATP in a process called chemiosmosis. ATP is generated by the removal of hydrogen ions from NADH and $FADH_2$. This yields 34 ATP molecules.

BIOLOGY

37. Which of these is a function of the cardiovascular system?
(Organismal Biology)

A. Move oxygenated blood around the body

B. Oxygenate the blood through gas exchange

C. Act as an exocrine system

D. Flush toxins out of the body

E. Transport signals from the brain

The answer is A.
The cardiovascular system moves oxygenated blood around the body via the heart (a pump) and tubes (arteries and veins).

38. Which of these is not a part of the nervous system?
(Organismal Biology)

A. brain

B. spinal cord

C. axons

D. venules

E. cochlea

The answer is D.
Venules are part of the circulatory system. The others are part of the nervous system.

BIOLOGY

39. Organisms need to maintain a constant internal environment to survive. Which of these is a method by which they achieve this?
(Organismal Biology)

   A. respiration

   B. reproduction

   C. depolarization

   D. repolarization

   E. thermoregulation

The answer is E.
Thermoregulation is how an organism maintains its body temperature. If it is an endothermic organism, it can respond to changes in temperature by sweating or growing more fur. If it is an ectothermic organism, it can move to a warmer or cooler location.

40. Which of these controls the body's endocrine mechanisms?
(Organismal Biology)

   A. feedback loops

   B. control molecules

   C. neurochemicals

   D. neurotransmitters

   E. behavioral responses

The answer is A.
The body's mechanisms are controlled by feedback loops.

BIOLOGY

41. What is the gland that regulates the calcium in the body?
(Organismal Biology)

   A. Thyroid gland

   B. Parathyroid gland

   C. Hypothalamus

   D. Pituitary gland

   E. Pancreas

The answer is B.
The parathyroid glands regulate the calcium levels in the body. They are imbedded within the thyroid gland.

42. Which of these steroids is not created in the gonads?
(Organismal Biology)

   A. Testosterone

   B. Estrogen

   C. Progesterone

   D. ACTH

   E. FSH

The answer is D.
ACTH is not one of the three steroids produced by the gonads. The other three are made by the gonads.

BIOLOGY

43. **What is the most common neurotransmitter?**
    (Organismal Biology)

    A. epinephrine

    B. serotonin

    C. acetyl choline

    D. norepinephrine

    E. oxytocin

The answer is C.
The most common neurotransmitter is acetyl choline.

44. **Food is carried through the digestive tract by a series of wave-like contractions. What is this process is called?**
    (Organismal Biology)

    A. peristalsis

    B. chyme

    C. digestion

    D. absorption

    E. depolarization

The answer is A.
Peristalsis is the process of wave-like contractions that moves food through the digestive tract.

BIOLOGY

45. Which of these must muscles pull on in order to initiate movement?
(Organismal Biology)

    A. skin

    B. bones

    C. joints

    D. ligaments

    E. bursa

The answer is B.
The muscular system's function is for movement. Skeletal muscles are attached to bones and are responsible for their movement.

46. Hormones are essential to the regulation of reproduction. What organ is responsible for the release of hormones for sexual maturity?
(Organismal Biology)

    A. pituitary gland

    B. hypothalamus

    C. pancreas

    D. thyroid gland

    E. pineal gland

The answer is B.
The hypothalamus begins secreting hormones that help mature the reproductive system and stimulate development of the secondary sex characteristics.

BIOLOGY

47. What is the type of muscle in the human body that is voluntary?
(Organismal Biology)

    A. Cardiac

    B. Sarcomere

    C. Smooth

    D. Skeletal

    E. Actin

The answer is D.
Of all of the above, only skeletal muscle is under voluntary control. It is found in the skeletal muscles of the human body.

48. The wrist is an example of what kind of joint?
(Organismal Biology)

    A. Ball and socket

    B. Pivot

    C. Stationary

    D. Hinge

    E. Gliding

The answer is B.
The wrist joint is a pivot joint.

BIOLOGY

49. **What is the waterproofing protein in the skin called?**
    (Organismal Biology)

    A. actin

    B. epidermis

    C. collagen

    D. sebum

    E. keratin

The answer is E.
The waterproofing protein in the skin is called keratin.

50. **What is the muscular adaptation called that is used to move food through the digestive system?**
    (Organismal Biology)

    A. peristalsis

    B. passive transport

    C. voluntary action

    D. bulk transport

    E. endocytosis

The answer is A.
Peristalsis is a process of wave-like contractions. This process allows food to be carried down the pharynx and though the digestive tract.

BIOLOGY

51. What is the role of neurotransmitters in nerve action?
(Organismal Biology)

    A. to turn off the sodium pump

    B. to turn off the calcium pump

    C. to send impulses to neurons

    D. to send impulses around the body

    E. to send impulses from axon to dendrite

The answer is A.
The neurotransmitters turn off the sodium pump, which results in depolarization of the membrane.

52. Fats are broken down by which substance?
(Organismal Biology)

    A. bile produced in the gall bladder

    B. lipase produced in the gall bladder

    C. glucagons produced in the liver

    D. amylase produces in the gall bladder

    E. bile produced in the liver

The answer is E.
The liver produces bile, which breaks down and emulsifies fatty acids.

BIOLOGY

53. Where does fertilization in humans usually occurs?
(Organismal Biology)

    A. uterus

    B. ovary

    C. fallopian tubes

    D. vagina

    E. epididymis

The answer is C.
Fertilization of the egg by the sperm normally occurs in the fallopian tube. The fertilized egg is then implanted on the uterine lining for development.

54. Which of these is lacking in the dermis layer of skin?
(Organismal Biology)

    A. sweat glands

    B. keratin

    C. hair follicles

    D. blood vessels

    E. living cells

The answer is B.
Keratin is a water proofing protein found in the epidermis.

BIOLOGY

55. A school age boy had the chicken pox as a baby. Why will he most likely not get this disease again?
    (Organismal Biology)

    A. passive immunity

    B. vaccination

    C. antibiotics

    D. active immunity

    E. antigen production

The answer is D.
Active immunity develops after recovery from an infectious disease, such as the chicken pox, or after vaccination. Passive immunity to some diseases may be passed from one individual to another (from mother to nursing child).

56. What is any foreign particle called that causes an immune reaction?
    (Organismal Biology)

    A. an antigen

    B. a histocompatibity complex

    C. an antibody

    D. a vaccine

    E. a bacteriophage

The answer is A.
An antigen is any foreign particle that results in an immune reaction.

BIOLOGY

57. Which of these statements describes the polymerase chain reaction?
(Population Biology)

    A. It is a group of polymerases.

    B. It is a technique for amplifying DNA.

    C. It is a primer for DNA synthesis.

    D. It is a way to synthesize polymerase.

    E. It is a series of genetic mutations.

The answer is B.
PCR is a technique in which a piece of DNA can be amplified into billions of copies within a few hours.

58. Which part of a DNA nucleotide can vary?
(Molecular & Cell Biology)

    A. deoxyribose

    B. phosphate group

    C. hydrogen bonds

    D. sugar

    E. nitrogenous base

The answer is C.
DNA is made of a 5-carbon sugar (deoxyribose), a phosphate group, and a nitrogenous base. There are four nitrogenous bases in DNA that vary to allow for the four different nucleotides.

BIOLOGY

59. A DNA strand has the base sequence of TCAGTA. Its DNA complement would have which of the following sequences?
(Molecular & Cell Biology)

    A. ATGACT

    B. TCAGTA

    C. AGUCAU

    D. AGTCAT

    E. TCTGTA

The answer is D.
The complement strand to a single strand DNA molecule has a complementary sequence to the template strand. T pairs with A and C pairs with G. Therefore, the complement to TCAGTA is AGTCAT.

60. Which of these carries amino acids to the ribosome during protein synthesis?
(Molecular & Cell Biology)

    A. messenger RNA

    B. ribosomal RNA

    C. transfer RNA

    D. DNA

    E. RNA

The answer is C.
The tRNA molecule is specific for a particular amino acid. The tRNA has an anticodon sequence that is complementary to the codon. This specifies where the tRNA places the amino acid in protein synthesis.

BIOLOGY

61. A protein is sixty amino acids in length. This requires a coded DNA sequence of how many nucleotides?
(Molecular & Cell Biology)

   A. 20

   B. 30

   C. 120

   D. 180

   E. 240

The answer is D.
Each amino acid codon consists of 3 nucleotides. If there are 60 amino acids in a protein, then 60 x 3 = 180 nucleotides.

62. A DNA molecule has the sequence of ACTATG. What is the anticodon of this molecule?
(Molecular & Cell Biology)

   A. UGAUAC

   B. ACTATG

   C. TGATAC

   D. ACUAUG

   E. CTGCGA

The answer is D.
The DNA is first transcribed into mRNA. Here, the DNA has the sequence ACTATG; therefore, the complementary mRNA sequence is UGAUAC (remember, in RNA, T is U). This mRNA sequence is the codon. The anticodon is the complement to the codon. The anticodon sequence will be ACUAUG (remember, the anticodon is tRNA, so U is present instead of T).

BIOLOGY

63. **What is the general term for a change that affects the sequence of bases in a gene?**
   (Molecular & Cell Biology)

   A. deletion

   B. polyploid

   C. mutation

   D. duplication

   E. substitution

The answer is C.
A mutation is an inheritable change in DNA. It may be an error in replication or a spontaneous rearrangement of one ore more segments of DNA. Deletion and duplication are types of mutations. Polyploidy is when an organism has more than two complete chromosome sets.

64. **Segments of DNA can be transferred from the DNA of one organism to another through the use of which of the following?**
   (Population Biology)

   A. bacterial plasmids

   B. viruses

   C. chromosomes from frogs

   D. plant DNA

   E. Okazaki fragments

The answer is A.
Plasmids can transfer themselves (and therefore their genetic information) by a process called conjugation. This requires cell-to-cell contact.

BIOLOGY

65. What is the enzyme that unwinds DNA during replication?
(Molecular & Cell Biology)

A. DNAse

B. DNA replicase

C. DNA helicase

D. DNA topoisomerases

E. DNA polymerase

The answer is C.
The enzyme helicase is involved in unwinding DNA during replication.

66. What is a small circular piece of DNA called that contains accessory DNA?
(Molecular & Cell Biology)

A. mitochondrial DNA

B. messenger RNA

C. transfer DNA

D. Okazaki fragment

E. plasmid

The answer is E.
A plasmid is a small, circular piece of accessory DNA.

BIOLOGY

67. In DNA, adenine bonds with _____, while cytosine bonds with _____.
(Molecular & Cell Biology)

A. thymine/guanine

B. adenine/cytosine

C. cytosine/uracil

D. guanine/thymine

E. uracil/adenine

The answer is A.
In DNA, adenine pairs with thymine and cytosine pairs with guanine because of their nitrogenous base structures.

68. Which protein structure consists of the coils and folds of polypeptide chains?
(Molecular & Cell Biology)

A. secondary structure

B. quaternary structure

C. tertiary structure

D. primary structure

E. quinary structure

The answer is A.
Primary structure is the protein's unique sequence of amino acids. Secondary structure is the coils and folds of polypeptide chains. The coils and folds are the result of hydrogen bonds along the polypeptide backbone. Tertiary structure is formed by bonding between the side chains of the amino acids. Quaternary structure is the overall structure of the protein from the aggregation of two or more polypeptide chain

BIOLOGY

69. What can be said about homozygous individuals?
(Organismal Biology)

    A. They have two different alleles.

    B. They are of the same species.

    C. They exhibit the same features.

    D. They have a pair of identical alleles.

    E. They produce identical offspring.

The answer is D.
Homozygous individuals have a pair of identical alleles and heterozygous individuals have two different alleles.

70. The term "phenotype" refers to which of the following?
(Organismal Biology)

    A. a condition that is heterozygous

    B. the genetic makeup of an individual

    C. a condition that is homozygous

    D. how the genotype is expressed

    E. from which parent the traits were inherited

The answer is D.
Phenotype is the physical appearance or expression of an organism due to its genetic makeup (genotype).

BIOLOGY

71. **The ratio of brown-eyed to blue-eyed children from the mating of a blue-eyed male to a heterozygous brown-eyed female is expected to be which of the following?**
    (Organismal Biology)

    A. 3:1

    B. 2:2

    C. 1:0

    D. 1:2

    E. 0:4

The answer is B.
Use a Punnet square to determine the ratio.

	b	b
B	Bb	Bb
b	bb	bb

B = brown eyes, b = blue eyes

Female genotype is on the side and the male genotype is across the top.

The female is heterozygous and her phenotype is brown eyes. This means the dominant allele is for brown eyes. The male expresses the homozygous recessive allele for blue eyes. Their children are expected to have a ratio of brown eyes to blue eyes of 2:2; or 1:1.

72. **Which of these defines the Law of Segregation defined by Gregor Mendel?**
    (Organismal Biology)

    A. After meiosis, each new cell will contain an allele that is recessive.

    B. Only one of two alleles is expressed in a heterozygous organism.

    C. The allele expressed is always the dominant allele.

    D. Alleles of one trait do not affect the inheritance of alleles on another chromosome.

    E. When sex cells form, the two alleles that determine a trait will end up on different gametes.

The answer is E.
The law of segregation states that the two alleles for each trait segregate into different gametes.

BIOLOGY

73. **Which of the following is an example of the incomplete dominance that occurs when a white flower is crossed with a red flower?**
    (Organismal Biology)

    A. pink flowers

    B. red flowers

    C. white flowers

    D. red and white flowers

    E. white and pink flowers

The answer is A.
Incomplete dominance is when the F_1 generation results in an appearance somewhere between the parents. Red flowers crossed with white flowers results in an F_1 generation with pink flowers.

74. **A child with type O blood has a father with type A blood and a mother with type B blood. The genotypes of the parents respectively would be which of the following?**
    (Organismal Biology)

    A. AA and BO

    B. AO and BO

    C. AA and BB

    D. AO and OO

    E. OO and AB

The answer is B.
Type O blood has 2 recessive O genes. A child receives one allele from each parent; therefore, each parent in this example must have an O allele. The father has type A blood with a genotype of AO and the mother has type B blood with a genotype of BO.

BIOLOGY

75. **Crossing over, which increases genetic diversity, occurs during which stage(s) of meiosis?**
    (Molecular & Cell Biology)

    A. telophase II in meiosis

    B. metaphase in mitosis

    C. interphase in meiosis

    D. prophase I in meiosis

    E. metaphase II in meiosis

The answer is D.
During prophase I of meiosis, the replicated chromosomes condense and pair with their homologues in a process called synapsis. Crossing over, the exchange of genetic material between homologues to further increase diversity, occurs during prophase I of meiosis.

76. **ABO blood grouping is an example of which type of allele dominance?**
    (Organismal Biology)

    A. Autosomal dominance

    B. Incomplete dominance

    C. Somatic dominance

    D. Complete dominance

    E. Codominance

The answer is E.
ABO blood grouping involves codominance. This means that more than one allele can express itself at the same time.

BIOLOGY

77. **In a Punnett square with a single trait, what are the ratios of genotypes produced between two heterozygous individuals?**
    (Organismal Biology)

    A. 1:2:2

    B. 2:1:1

    C. 1:1:1

    D. 1:2:1

    E. 2:2:2

The answer is D.
The Punnet square ratio for a single trait is 1:2:1. All three possible genotypes will be expressed – homozygous dominant, heterozygous, and homozygous recessive.

78. **What is the term for an organism's genetic makeup?**
    (Organismal Biology)

    A. Heterozygote

    B. Genotype

    C. Phenotype

    D. Homozygote

    E. Dominance

The answer is B.
The genetic makeup is called the genotype.

BIOLOGY

79. Which of these represents a genetic engineering advancement in the medical field?
(Population Biology)

    A. stem cell reproduction

    B. pesticides

    C. degradation of harmful chemicals

    D. antibiotics

    E. gene therapy

The answer is E.
Gene therapy is the introduction of a normal allele to the somatic cells to replace a defective allele. The medical field has had success in treating patients with a single enzyme deficiency disease. Gene therapy has allowed doctors and scientists to introduce a normal allele that provides the missing enzyme.

80. Which of the following is not true regarding restriction enzymes?
(Population Biology)

    A. They aid in transcombination procedures.

    B. They are used in genetic engineering.

    C. They are named after the bacteria in which they naturally occur.

    D. They identify and splice certain base sequences on DNA.

    E. They can be produced by certain lipids during DNA replication.

The answer is A.
A restriction enzyme is a bacterial enzyme that cuts foreign DNA at specific locations. The splicing of restriction fragments into a plasmid results in a recombinant plasmid.

BIOLOGY

81. **Which of these processes is not one of the modern uses of DNA?**
    (Population Biology)

    A. PCR technology

    B. Gene therapy

    C. Cloning

    D. Genetic Alignment

    E. Transgenic organisms

The answer is D.
PCR technology, gene therapy and cloning all come out of working with DNA.

82. **Which statement best represents gel electrophoresis?**
    (Population Biology)

    A. It isolates fragments of DNA for scientific purposes.

    B. It cannot be used in proteins.

    C. It requires the polymerase chain reaction.

    D. It only separates DNA by size.

    E. It uses different charged particles to color the bands.

The answer is A.
Gel electrophoresis separates DNA by size and charge. It can be used in proteins as well and is not dependent on the polymerase chain reaction.

BIOLOGY

83. **What is the term that describes the duplication of genetic material into another cell?**
    (Population Biology)

    A. replicating

    B. cell duplication

    C. transgenics

    D. genetic restructuring

    E. cloning

The answer is E.
Cloning is the duplication of genetic material into another cell.

84. **What does gel electrophoresis use to separate the DNA?**
    (Population Biology)

    A. the amount of current

    B. the size of the molecule

    C. the positive charge of the molecule

    D. the solubility of the gel

    E. the source of the DNA

The answer is B.
Electrophoresis uses electrical charges of molecules to separate them according to their size.

BIOLOGY

85. Which of these is a result of reproductive isolation?
(Population Biology)

- A. extinction
- B. migration
- C. fossilization
- D. speciation
- E. radiation

The answer is D.
Reproductive isolation is caused by any factor that impedes two species from producing viable, fertile hybrids. Reproductive isolation of populations is the primary criterion for recognition of species status.

86. Which of these is true about natural selection?
(Population Biology)

- A. It acts on an individual genotype.
- B. It is not currently happening.
- C. It is only an animal phenomenon.
- D. It acts on the individual phenotype.
- E. It is used to prevent overpopulation.

The answer is D.
Natural selection acts on the individual phenotype.

BIOLOGY

87. How does diversity aid a population?
(Population Biology)

    A. Individuals are better able to survive.

    B. Mates are attracted to a diverse population.

    C. Potential mates like conformity.

    D. It increases the DNA differences in the population.

    E. It provides possible improvements to the population.

The answer is E.
Diversity provides possible improvements to the population.

88. Which statement is not true about diversity?
(Population Biology)

    A. Without diversity there would be extinction.

    B. Diversity is increasing all the time.

    C. Fossil evidence supports diversity.

    D. Sexual reproduction encourages more diversity.

    E. Skeletons are too similar to allow for diversity.

The answer is E.
The other answers are all true. Without diversity, there would be extinction, diversity is increasing all the time and fossil evidence supports an increase in diversity.

BIOLOGY

89. Which of these ideas was a major part of Darwin's evolutionary theory?
(Population Biology)

    A. Punctualism

    B. Gradualism

    C. Equilibrium

    D. Convergency

    E. Altruism

The answer is B.
Darwin's book is based upon gradualism, the idea that species change slowly over time.

90. Which statement is not true about reproductive isolation?
(Population Biology)

    A. It prevents populations from exchanging genes.

    B. It can occur by preventing fertilization.

    C. It can result in speciation.

    D. It happens more often on the mainland.

    E. It produces offspring with unique phenotypes

The answer is D.
Reproductive isolation can result in speciation, can occur by preventing fertilization and prevents populations from exchanging genes. It is a common phenomenon of islands.

BIOLOGY

91. Which idea is true about members of the same species?
(Population Biology)

    A. They look identical.

    B. They never change.

    C. They reproduce successfully within their group.

    D. They live in the same geographic location.

    E. They have very dissimilar genotypes.

The answer is C.
Species are defined by the ability to successfully reproduce with members of their own kind.

92. Which of the following factors will affect the Hardy-Weinberg law of equilibrium, leading to evolutionary change?
(Population Biology)

    A. no mutations

    B. non-random mating

    C. no immigration or emigration

    D. large population

    E. small individual species

The answer is B.
There are five requirements to keep the Hardy-Weinberg equilibrium stable: no mutation, no selection pressures, an isolated population, a large population, and random mating.

BIOLOGY

93. **If a population is in Hardy-Weinberg equilibrium and the frequency of the recessive allele is 0.3, what percentage of the population is expected to be heterozygous?**
(Population Biology)

   A. 9%

   B. 49%

   C. 42%

   D. 21%

   E. 7%

The answer is C.
0.3 is the value of q. Therefore, $q^2 = 0.09$. According to the Hardy-Weinberg equation, $1 = p + q$.

$1 = p + 0.3$.
$p = 0.7$
$p^2 = 0.49$ Next, plug q^2 and p^2 into the equation $1 = p^2 + 2pq + q^2$.

$1 = 0.49 + 2pq + 0.09$ (where 2pq is the number of heterozygotes).
$1 = 0.58 + 2pq$
$2pq = 0.42$ Multiply by 100 to get the percent of heterozygotes, 42%.

94. **Which aspect of science does not support evolution?**
(Population Biology)

   A. comparative anatomy

   B. organic chemistry

   C. comparison of DNA among organisms

   D. analogous structures

   E. embryology

The answer is B.
Comparative anatomy is the comparison of characteristics of the anatomies of different species. This includes homologous structures and analogous structures. The comparison of DNA between species is the best known way to place species on the evolution tree. Organic chemistry has nothing to do with evolution.

BIOLOGY

95. In which of these does evolution occurs?
(Population Biology)

    A. individuals

    B. populations

    C. organ systems

    D. cells

    E. ecosystems

The answer is B.
Evolution is a change in genotype over time. Gene frequencies shift and change from generation to generation. Populations evolve, not individuals.

96. Which process contributes most to the large variety of living things in the world today?
(Population Biology)

    A. meiosis

    B. asexual reproduction

    C. mitosis

    D. alternation of generations

    E. reproductive isolation

The answer is A.
During meiosis prophase I crossing over occurs. This exchange of genetic material between homologues increases diversity.

BIOLOGY

97. **Which of the following gases was a major part of the primitive Earth atmosphere?**
    (Population Biology)

    A. fluorine

    B. methane

    C. oxygen

    D. krypton

    E. argon

The answer is B.
The primitive atmosphere contained ammonia, methane and hydrogen but very little oxygen.

98. **What is a major principle of the Endosymbiotic Theory?**
    (Population Biology)

    A. Birds and dinosaurs share a common ancestor.

    B. Animals evolved in close relationships with one another.

    C. Prokaryotes arose from eukaryotes.

    D. Inorganic compounds are the basis of living things.

    E. Eukaryotes arose from very simple prokaryotes.

The answer is E.
The Endosymbiotic theory of the origin of eukaryotes states that eukaryotes arose from symbiotic groups of prokaryotic cells. According to this theory, smaller prokaryotes lived within larger prokaryotic cells, eventually evolving into chloroplasts and mitochondria.

BIOLOGY

99. The wing of a bird, the human arm, and the pectoral fluke of a whale all have the same bone structure. What are these structures called?
(Population Biology)

    A. polymorphic structures

    B. homologous structures

    C. vestigial structures

    D. analogous structures

    E. allopatric structures

The answer is B.
Homologous structures have the same genetic basis (leading to similar appearances), but are used for different functions.

100. Which of the following is not an abiotic factor?
(Population Biology)

    A. temperature

    B. rainfall

    C. soil quality

    D. predation

    E. wind speed

The answer is D.
Abiotic factors are non-living aspects of an ecosystem. Temperature, rainfall, and soil quality are all abiotic factors. Predation is an example of a biotic factor-- living things.

BIOLOGY

101. What is not true about cladistics?
(Population Biology)

A. It is the study of phylogenetic relationships of organisms.

B. It involves a branching diagram that uses the development of novel traits to separate groups of organisms.

C. It distinguishes between the relative importance of the traits.

D. It shows when traits developed with respect to other traits.

E. It indicates which organisms are most closely related to each other and what their common ancestors were.

The answer is C.
Cladistics does not show how important certain traits were to different species. It represents when species evolved and how closely related they are to each other.

102. If DDT were present in an ecosystem, which of the following organisms would have the highest concentration in its body?
(Population Biology)

A. herring

B. diatom

C. zooplankton

D. salmon

E. osprey

The answer is E.
Chemicals and pesticides accumulate along the food chain. Tertiary consumers have more accumulated toxins than animals at the bottom of the food chain.

BIOLOGY

103. What eats secondary consumers?
(Population Biology)

    A. Producers

    B. Tertiary consumers

    C. Primary consumers

    D. Decomposers

    E. Detritivores

The answer is B.
The tertiary consumers eat the secondary consumers and the secondary consumers eat the primary consumers.

104. Which statement is true about the water cycle?
(Population Biology)

    A. Two percent of the water is fixed and unavailable.

    B. 75% of available water is groundwater.

    C. The water cycle is driven by the ocean currents.

    D. Surface water percolates up from underground springs.

    E. New water is being added into the cycle all the time.

The answer is A.
96 percent of available water is groundwater. The water cycle is driven by the sun. Surface water is available.

BIOLOGY

105. Which statement about the carbon cycle is false?
(Population Biology)

　　A. Ten percent of all available carbon is in the air.

　　B. Carbon dioxide is fixed by glycosylation.

　　C. Plants fix carbon in the form of glucose.

　　D. Animals release carbon through respiration.

　　E. Most atmospheric carbon comes from the decay of dead organisms.

The answer is B.
Ten percent of all available carbon is in the air. Plants fix carbon via photosynthesis to make glucose and animals release carbon through respiration.

106. What is the impact of sulfur oxides and nitrogen oxides in the environment when they react with water?
(Population Biology)

　　A. ammonia

　　B. acidic precipitation

　　C. sulfuric acid

　　D. global warming

　　E. greenhouse effect

The answer is B.
Acidic precipitation is rain, snow, or fog with a pH less than 5.6. It is caused by sulfur oxides and nitrogen oxides that react with water in the air to form acids that fall down to Earth as precipitation.

BIOLOGY

107. Which term is not associated with the water cycle?
(Population Biology)

    A. precipitation

    B. transpiration

    C. fixation

    D. evaporation

    E. runoff

The answer is C.
Water is recycled through the processes of evaporation and precipitation. Transpiration is the evaporation of water from leaves. Fixation is not associated with the water cycle.

108. Which of the following is a density dependent factor that affects a population?
(Population Biology)

    A. temperature

    B. rainfall

    C. predation

    D. soil nutrients

    E. wind speed

The answer is C.
As a population increases, the competition for resources is intense and the growth rate declines. This is a density-dependent factor. An example of this would be predation. Density-independent factors affect the population regardless of its size. Examples of density-independent factors are rainfall, temperature, and soil nutrients.

BIOLOGY

109. High humidity and temperature stability are present in which of the following biomes?
(Population Biology)

A. taiga

B. deciduous forest

C. desert

D. tropical rain forest

E. coniferous forest

The answer is D.
A tropical rain forest is located near the equator. Its temperature is at a constant 25 degrees C and the humidity is high due to the rainfall that exceeds 200 cm per year.

110. Which trophic level has the highest ecological efficiency?
(Population Biology)

A. decomposers

B. producers

C. tertiary consumers

D. secondary consumers

E. primary consumers

The answer is B.
The amount of energy that is transferred between trophic levels is called the ecological efficiency. The visual of this is represented in a pyramid of productivity. The producers have the greatest amount of energy and are at the bottom of this pyramid.

BIOLOGY

111. From where does the oxygen created in photosynthesis come?
(Molecular & Cell Biology)

    A. carbon dioxide

    B. chlorophyll

    C. glucose

    D. carbon monoxide

    E. water

The answer is E.
In photosynthesis, water is split; the hydrogen atoms are pulled to carbon dioxide that is taken in by the plant and ultimately reduced to make glucose. The oxygen from the water is given off as a waste product.

112. Which of the following is true of decomposers?
(Population Biology)

    A. Decomposers recycle the carbon accumulated in durable organic material.

    B. They take nitrogen out of the soil to use for food.

    C. Decomposers absorb nutrients from the air to maintain their metabolisms.

    D. Decomposers belong to the Genus *Escherichia*.

    E. They are able to use the Sun to produce their own energy.

The answer is A.
Decomposers recycle phosphorus and carbon and undergo ammonification. The break down dead organisms to release the carbon held within their tissues. This carbon then reenters the ecosystem.

BIOLOGY

113. A clownfish is protected by a sea anemone's tentacles, and in turn, the anemone receives uneaten food from the clownfish. What type of symbiosis is exemplified by this example?
    (Population Biology)

    A. mutualism

    B. parasitism

    C. commensalism

    D. competition

    E. amensalism

The answer is A.
Neither the clownfish nor the anemone cause harmful effects towards one another and they both benefit from their relationship. Mutualism is when two species that occupy a similar space benefit from their relationship.

114. Which of these is most likely to happen in order for primary succession to occur?
    (Population Biology)

    A. nutrient enrichment

    B. a forest fire

    C. bare rock is exposed after a water table recedes

    D. a housing development is built

    E. a farmer stops cultivating her fields

The answer is C.
Primary succession occurs where life never existed before, such as flooded areas or a new volcanic island. It is only after the water recedes that the rock is able to support new life.

BIOLOGY

115. What is the Mendelian law called that states that only one of the two possible alleles from each parent is passed on to the offspring?
(Organismal Biology)

    A. The Mendelian Law

    B. The Law of Independent Assortment

    C. The Law of Segregation

    D. The Allele Law

    E. The Law of Dominance and Recessiveness

The answer is B.
The law of independent assortment states that only one of a pair of alleles is transferred from parent to offspring.

CHEMISTRY

Description of the Examination

The Chemistry examination covers material that is usually taught in a one-year college course in general chemistry. Understanding of the structure and states of matter, reaction types, equations and stoichiometry, equilibrium, kinetics, thermodynamics, and descriptive and experimental chemistry is required, as is the ability to interpret and apply this material to new and unfamiliar problems. During this examination, an online scientific calculator and a periodic table are available as part of the testing software.

The examination contains approximately 75 questions to be answered in 90 minutes. Some of these are pretest questions that will not be scored. Any time spent on tutorials and providing personal information is in addition to the actual testing time.

Knowledge and Skills Required

Questions on the Chemistry examination require candidates to demonstrate one or more of the following abilities.

- Recall - remember specific facts; demonstrate straightforward knowledge of information and familiarity with terminology.
- Application - understand concepts and reformulate information into other equivalent terms; apply knowledge to unfamiliar and/or practical situations; use of mathematics to solve chemistry problems.
- Interpretation - infer and deduce from data available and integrate information to draw conclusions, and recognize unstated assumptions.

The subject matter of the Chemistry examination is drawn from the following topics. The percentages next to the main topics indicate the approximate percentage of exam questions on that topic.

Scientific Calculator

A scientific (nongraphing) calculator is integrated into the exam software, and it is available to students during the entire testing time. Students are expected to know how and when to make appropriate use of the calculator. The scientific calculator for the iBT versions of the CLEP exams, together with a brief video tutorial, is available to students as a free download for a 30-day trial period. Students are encouraged to download the calculator and become familiar with its functionality prior to taking the exam.

Students will find the online scientific calculator helpful in performing calculations (e.g., arithmetic, exponents, roots, logarithms).

The eCBT and iBT versions of the scientific calculators look different, but both have the necessary functions that will help the students to answer questions during the exams.

20% Structure of Matter
Atomic theory and atomic structure
- Basics of the atomic theory.
- Atomic masses; determination by chemical and physical means.
- Atomic numbers and mass numbers; isotopes and mass spectroscopy.
- Electron energy levels: atomic spectra, quantum numbers, and atomic orbitals.
- Periodic relationships, including, for example, atomic radii, ionization energies, electron affinities, and oxidation states.

CHEMISTRY

Chemical bonding
- Binding forces
  - Types: covalent, ionic, metallic, macromolecular (or network), dispersion, and hydrogen bonding.
  - Relationships to structure and to properties.
  - Polarity of bonds; electronegativities.
- Geometry of molecules, ions, and coordination complexes: structural isomerism, dipole moments of molecules, and relation of properties to structure.
- Molecular models
  - Valence bond theory, hybridization of orbitals, resonance, and sigma and pi bonds.
  - Other models such as molecular orbitals.
- Nuclear chemistry: nuclear equations, half-lives, and radioactivity; and chemical applications.

19% States of Matter

Gases
- Laws of ideal gases; equations of state for an ideal gas.
- Kinetic-molecular theory:
  - Interpretation of ideal gas laws on the basis of this theory.
  - The mole concept; Avogadro's number.
  - Dependence of kinetic energy of molecules on temperature: Boltzmann distribution.
  - Deviations from ideal gas laws.

Liquids and solids
- Liquids and solids from the kinetic molecular viewpoint.
- Phase diagrams of one-component systems.
- Changes of state, and critical phenomena.
- Crystal structure.

Solutions
- Types of solutions and factors affecting solubility.
- Methods of expressing concentration.
- Colligative properties; for example, Raoult's law.
- Effect of interionic attraction on colligative properties and solubility.

12% Reaction Types

Formation and cleavage of covalent bonds
- Acid-base reactions; concepts of Arrhenius, Brønsted-Lowry, and Lewis; amphoterism.
- Reactions involving coordination complexes.
- Precipitation reactions.

Oxidation-reduction reactions
- Oxidation number.
- The role of the electron in oxidation-reduction.
- Electrochemistry; electrolytic cells, standard half-cell potentials, prediction of the direction of redox reactions, and effect of concentration changes.

10% Equations and Stoichiometry
- Ionic and molecular species present in chemical systems; net-ionic equations.
- Stoichiometry: mass and volume relations with emphasis on the mole concept.
- Balancing of chemical reactions, including those for redox reactions.

CHEMISTRY

7% Equilibrium
Concept of dynamic equilibrium, physical and chemical; Le Châtelier's principle; equilibrium constants.

Quantitative treatment
- Equilibrium constants for gaseous reactions in terms of both molar concentrations and partial pressure (K_c, K_p).
- Equilibrium constants for reactions in solutions:
  - Constants for acids and bases; pK; pH.
  - Solubility-product constants and their application to precipitation and dissolution of slightly soluble compounds.
  - Constants for complex ions.
  - Common ion effect and buffers.

4% Kinetics
- Concept of rate of reaction.
- Order of reaction and rate constant: their determination from experimental data.
- Effect of temperature change on rate constants.
- Energy of activation; the role of catalysts.
- The relationship between the rate-determining step and a mechanism.

5% Thermodynamics
State functions
- First law: heat of formation; heat of reaction; change in enthalpy, Hess's law; heat capacity; heats of vaporization and fusion.
- Second law: free energy of formation; free energy of reaction; dependence of change in free energy on enthalpy and entropy changes.
- Relationship of change in free energy to equilibrium constants and electrode potentials.

14% Descriptive Chemistry
The accumulation of certain specific facts of chemistry is essential to enable students to comprehend the development of principles and concepts, to demonstrate applications of principles, to relate fact to theory and properties to structure, and to develop an understanding of systematic nomenclature that facilitates communication.

The following areas are normally included on the examination:
- Chemical reactivity and products of chemical reactions.
- Relationships in the periodic table: horizontal, vertical, and diagonal.
- Chemistry of the main groups and transition elements, including typical examples of each.
- Organic chemistry, including topics such as functional groups and isomerism (may be treated as a separate unit or as exemplary material in other areas, such as bonding).

9% Experimental Chemistry
Some experiments are based on laboratory practical work widely performed in general chemistry and ask about the equipment used, observations made, calculations performed, and interpretation of the results. The questions are designed to provide a measure of understanding of the basic tools of chemistry and their applications to simple chemical systems.

CHEMISTRY

SAMPLE TEST

DIRECTIONS: Read each item and select the best response.

1. A piston compresses a gas at constant temperature. Which gas properties increase?

   I. Average speed of molecules
   II. Pressure
   III. Molecular collisions with container walls per second

   A. I and II

   B. I and III

   C. II and III

   D. I, II, and III

   E. None of the above

2. The temperature of a liquid is raised at atmospheric pressure. Which property of liquids increases?

   A. Critical pressure

   B. Vapor pressure

   C. Surface tension

   D. Viscosity

   E. Boiling Point

3. Potassium crystallizes with two atoms contained in each unit cell. What is the mass of potassium found in a lattice 1.00×10^6 unit cells wide, 2.00×10^6 unit cells high, and 5.00×10^5 unit cells deep?

   A. 85.0 µg

   B. 32.5 µg

   C. 64.9 µg

   D. 130 µg

   E. 130×10^6 µg

4. A gas is heated in a sealed container. Which of the following occur(s)?

   A. Gas pressure rises

   B. Gas density decreases

   C. The average distance between molecules increases

   D. The volume increases

   E. All of the above

CHEMISTRY

5. How many molecules are in 2.20 pg of a protein with a molecular weight of 150 kDa?

   A. 8.83×10^9

   B. 1.82×10^9

   C. 8.83×10^6

   D. 1.82×10^6

   E. 8.83×10^{15}

6. At STP, 20 µL of O_2 contain 5.4×10^{16} molecules. According to Avogadro's hypothesis, how many molecules are in 20 µL of Ne?

   A. 5.4×10^{15}

   B. 1.0×10^{16}

   C. 2.7×10^{16}

   D. 5.4×10^{16}

   E. 1.3×10^6

7. An ideal gas at 50.0 °C and 3.00 atm is enclosed in a 300 cm³ cylinder. The cylinder volume changes by moving a piston until the gas reaches 50.0 °C and 1.00 atm. What is the final volume?

   A. 100 cm³

   B. 450 cm³

   C. 900 cm³

   D. 1.20 dm³

   E. 150.0 cm³

8. 81-butanol, ethanol, methanol, and 1-propanol are all liquids at room temperature. Rank them in order of increasing viscosity.

   A. 1-butanol < 1-propanol < ethanol < methanol

   B. methanol < ethanol < 1-propanol < 1-butanol

   C. methanol < ethanol < 1-butanol < 1-propanol

   D. 1-propanol < 1-butanol < ethanol < methanol

   E. ethanol < methanol < 1-butanol. 1-propanol

9. One mole of an ideal gas at STP occupies 22.4 L. At what temperature will one mole of an ideal gas at 1 atm occupy 31.0 L?

   A. 34.6 °C

   B. 105 °C

   C. 378 °C

   D. 442 °C

   E. 28 °C

CHEMISTRY

10. What pressure is exerted by a mixture of 2.7 g of H_2 and 59 g of Xe at STP in a 50 L container?

    A. 0.69 atm
    B. 0.76 atm
    C. 0.80 atm
    D. 0.97 atm
    E. 27.0 atm

11. The normal boiling point of water on the Kelvin scale is closest to:

    A. 112 °K
    B. 212 °K
    C. 273 °K
    D. 373 °K
    E. 298 °K

12. Which phases may be present at the triple point of a substance?

    I. Gas
    II. Liquid
    III. Solid
    IV. Supercritical fluid

    A I, II, and III
    B. I, II, and IV
    C. II, III, and IV
    D. I, II, III, and IV
    E I, III, IV

13. The solubility of pure CO_2 in water at 25 °C and 1.0 atm is 0.034 M. According to Henry's law, what is the solubility of pure CO_2 in water at 25 °C and 4.0 atm? Assume no chemical reactions occur between CO_2 and H_2O.

    A. 0.0085 M
    B. 0.034 M
    C. 0.14 M
    D. 0.25 M
    E. 0.15 M

14. Which statement about molecular structures is false?

    A. [structure of 1,3-butadiene] is a conjugated molecule.

    B. A bonding σ orbital connects two atoms by a straight line between them.

    C. A bonding π orbital connects two atoms in a separate region from the straight line between them.

    D. [formate anion resonance structures] The anion with resonance forms always exists in one form or the other.

    E. They are all false

CHEMISTRY

15. What is the chemical composition of magnesium nitrate?

    A. Mg (11.1%), N (22.2%), O (66.7%)

    B. Mg (50.1%), N (22.2%), O (33.0%)

    C. Mg (16.4%), N (18.9%), O (64.7%)

    D. Mg (20.9%), N (24.1%), O (55.0%)

    E. Mg (28.2%), N (16.2%), O (55.7%)

16. How many neutrons are there in $^{60}_{27}Co$?

    A. 27

    B. 33

    C. 60

    D. 87

    E. 14

17. Select the list of atoms that is **arranged in order of increasing size**.

    A. Mg, Na, Si, Cl

    B. Si, Cl, Mg, Na

    C. Cl, Si, Mg, Na

    D. Na, Mg, Si, Cl

    E. Mg, Si, Cl, Na

18. Based on trends in the Periodic Table, which of the following properties would you expect to be greater for Rb than for K?

    I. Density
    II. Melting point
    III. Ionization energy
    IV. Oxidation number in a compound with chlorine

    A. I only

    B. I, II, and III

    C. II and III

    D. I, II, III, and IV

    E. None of the above

19. Why does $CaCl_2$ have a higher normal melting point than NH_3?

    A. London dispersion forces in $CaCl_2$ are stronger than covalent bonds in NH_3.

    B. Covalent bonds in NH_3 are stronger than dipole-dipole bonds in $CaCl_2$.

    C. Ionic bonds in $CaCl_2$ are stronger than London dispersion forces in NH_3.

    D. Ionic bonds in $CaCl_2$ are stronger than hydrogen bonds in NH_3.

    E. None of the Above

CHEMISTRY

20. **Rank the following bonds from least to most polar:**

    C-H, C-Cl, H-H, C-F

    A. C-H < H-H < C-F < C-Cl

    B. H-H < C-H < C-F < C-Cl

    C. C-F < C-Cl < C-H < H-H

    D. H-H < C-H < C-Cl < C-F

    E. H-H < C-H < C-Cl < C-F

21. **In C_2H_2, each carbon atom contains the following valence orbitals:**

    A. p only

    B. p and sp hybrids

    C. p and sp^2 hybrids

    D. sp^3 hybrids only

    E. s, p, and sp^2 hybrids

22. **The boiling points of N_2, O_2 and Cl_2 are respectively -196, -182, and -34 °C. Cl_2 boils at a much higher temperature than expected. This is explained by:**

    A. Single bonds are easier to break than double or triple bonds.

    B. Cl_2 has a longer bond length than the others.

    C. Cl_2 has greater London Dispersion forces because it has more electrons.

    D. Hydrogen bonding is stronger in Cl_2.

    E. None of the above.

23. **Which of the following statements regarding molecular geometries are true?**

    I. Double bonds are planar
    II. A central atom in sp^2 hybridization is trigonal
    III. A central atom with sp^2 hybridization of bipyramidal
    IV. Single bonds are always planar

    A. I and II

    B. I and IV

    C. II and IV

    D. II and III

    E. I, II, and IV

CHEMISTRY

24. The electronic configuration of Iron (Fe) is:

    A. $1s^2\ 2s^2\ 2p^6\ 3s^2\ 3p^6\ 4s^2\ 3d^6$

    B. $1s^2\ 2s^2\ 2p^6\ 3s^2\ 3p^6\ 3d^8$

    C. $1s^2\ 2s^2\ 2p^6\ 3s^2\ 3p^6\ 4s^2$

    D. $1s^2\ 2s^2\ 2p^6\ 3s^2\ 3p^6\ 4s^2\ 4p^6$

    E. $1s^2\ 2s^2\ 2p^6\ 3s^2\ 3p^6\ 4s^2\ 4p^3\ 4d^3$

25. Which of the following is a correct electron arrangement for oxygen?

    A.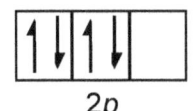
       1s   2s   2p

    B. $1s^2 1p^2 2s^2 2p^2$

    C. 2, 2, 4

    D. $2, 2, 4, \frac{1}{2}$

    E. None of the above

26. Which of the following is a proper Lewis dot structure of CHClO?

    A.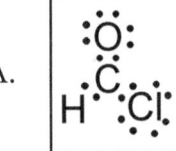

    B.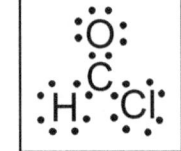

    C.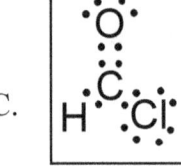

    D.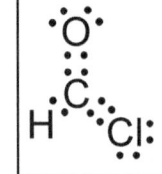

    E. None of the above

CHEMISTRY

27. Which intermolecular attraction force explains the following trend in straight-chain alkanes?

Condensed structural formula	Boiling point (°C)
CH_4	-161.5
CH_3CH_3	-88.6
$CH_3CH_2CH_3$	-42.1
$CH_3CH_2CH_2CH_3$	-0.5
$CH_3CH_2CH_2CH_2CH_3$	36.0
$CH_3CH_2CH_2CH_2CH_2CH_3$	68.7

A. London dispersion forces

B. Hydrophobic interactions

C. Dipole-dipole interactions

D. Hydrogen bonding

E. Ion-induced dipole interactions

28. Match the theory with the scientist who first proposed it:

I. Electrons, atoms, and all objects with momentum also exist as waves.
II. Electron density may be accurately described by a single mathematical equation.
III. There is an inherent indeterminacy in the position and momentum of particles.
IV. Radiant energy is transferred between particles in exact multiples of a discrete unit.

A. I - de Broglie, II - Planck, III - Schrödinger, IV - Thomson

B. I - Dalton, II - Bohr, III - Planck, IV - de Broglie

C. I - Henry, II - Bohr, III - Heisenberg, IV - Schrödinger

D. I - de Broglie, II - Schrödinger, III - Heisenberg, IV - Planck

E. I - Schrödinger, II - de Broglie, III - Plank, IV - Heisenberg

CHEMISTRY

29. The terrestrial composition of an element is: 50.7% as a stable isotope with an atomic mass of 78.9 μ and 49.3% as a stable isotope with an atomic mass of 80.9 μ. Calculate the atomic mass of the element.

    A. 79.0 μ

    B. 79.8 μ

    C. 79.9 μ

    D. 80.8 μ

    E. 80.0 μ

30. $^{3}_{1}H$ decays with a half-life of 12 years. 3.0 g of pure $^{3}_{1}H$ were placed in a sealed container 24 years ago. How many grams of $^{3}_{1}H$ remain?

    A. 0.38 g

    B. 0.75 g

    C. 1.5 g

    D. 0.125 g

    E. 3.0 g

31. Moving down a column on the Periodic Table:

    I. The atomic radius increases
    II. Ionization energy increases
    III. Protons are added
    IV. Metallic characteristics increase

    A. I, II, III

    B. I and II only

    C III only

    D. III and IV only

    E. None of the above

32. Moving from left to right on a Periodic Table (Li – to Ne) which of the following statements are true:

    I. The atomic radius decreases
    II. Electrons are added
    III. Ionization energies increases
    IV. Electronegativity decreases

    A. I only

    B. I and II

    C. I, II, and III

    D. I, II, III, and IV

    E. None of the above

CHEMISTRY

33. NH₄F is dissolved in water. Which of the following are the conjugate acid/base pairs present in the solution?

   I. NH_4^+/NH_4OH
   II. HF/F^-
   III. H_3O^+/H_2O
   IV. H_2O/OH^-

   A. I only

   B. I, II, and III

   C. I, III, and IV

   D. II and IV

   E. II, III, and IV

34. Rank the following from lowest to highest pH. Assume a small volume for the component given in moles:

   I. 0.01 mol HCl added to 1 L H₂O
   II. 0.01 mol HI added to 1 L of an acetic acid/sodium acetate solution at pH 4.0
   III. 0.01 mol NH₃ added to 1 L H₂O
   IV. 0.1 mol HNO₃ added to 1 L of a 0.1 M Ca(OH)₂ solution

   A. I < II < III < IV

   B. I < II < IV < III

   C. II < I < III < IV

   D. II < I < IV < III

   E. IV < III < II < I

35. Which statement about acids and bases is <u>not</u> true?

   A. All strong acids ionize in water.

   B. All Lewis acids accept an electron pair.

   C. All Brønsted bases use OH⁻ as a proton acceptor.

   D. All Arrhenius acids form H⁺ ions in water.

   E. Water can act as either an acid or a base.

36. What is the pH of a buffer solution made of 0.128 M sodium formate (HCOONa) and 0.072 M formic acid (HCOOH)? The pK_a of formic acid is 3.75.

   A. 2.0

   B. 3.0

   C. 3.75

   D. 4.0

   E. 5.0

CHEMISTRY

37. A 100 L vessel of pure O_2 at 500 kPa and 20 °C is used for the combustion of butane following the reaction:
$2C_4H_{10} + 13O_2 \rightarrow 8CO_2 + 10H_2O$

Find the mass of butane that should consume all the O_2 in the vessel. Assume O_2 is an ideal gas and use a value of $R = 8.314$ J/(mol·K).

A. 183 g

B. 467 g

C. 1.83 kg

D. 2.6 kg

E. 7.75 kg

38. 32.0 g of hydrogen and 32.0 grams of oxygen react together to form water until the limiting reagent is consumed. What reagents and products are present in the vessel after the reaction is complete?

A. 16.0 g O_2 and 48.0 g H_2O

B. 24.0 g H_2 and 40.0 g H_2O

C. 28.0 g H_2 and 36.0 g H_2O

D. 28.0 g H_2 and 34.0 g H_2O

E. 28.0 g H_2 and 16.0 g O_2

39. Which reaction is not a redox process?

A. Combustion of octane:
$2C_8H_{18} + 25O_2 \rightarrow 16CO_2 + 18H_2O$

B. Depletion of a lithium battery:
$Li + MnO_2 \rightarrow LiMnO_2$

C. Corrosion of aluminum by acid:
$2Al + 6HCl \rightarrow 2AlCl_3 + 3H_2$

D. Taking an antacid for heartburn:
$CaCO_3 + 2HCl \rightarrow CaCl_2 + H_2CO_3$
$\rightarrow CaCl_2 + CO_2 + H_2O$

E. None of the above

40. Which of the following are true about Galvanic cells?

I. Two half reactions take place in separate two chambers
II. Oxidation takes place at the anode
III. Reduction takes place at the cathode
IV. If the salt bridge is removed, the voltage drops to zero
V. Le Chatelier's principle can be applied to the systematic

A. I, II, and III

B. only II and III

C. only I and IV

D. All of the above

E. None of the above

CHEMISTRY

41. Balance the equation of the neutralization reaction between phosphoric acid and calcium hydroxide by filling in the blank stoichiometric coefficients from left to right.
    __H_3PO_4 + __$Ca(OH)_2$ → __$Ca_3(PO_4)_2$ + __H_2O

    A. 4, 3, 1, 4

    B. 2, 3, 1, 8

    C. 2, 3, 1, 6

    D. 2, 1, 1, 2

    E. 4, 3, 1, 1

42. Which of the following show the reaction between calcium nitrate and lithium sulfate in aqueous solution? Include all reagents and products, and make sure the reaction is mass balanced.

    A. $CaNO_3$ (aq) + Li_2SO_4 (aq) → $CaSO_4$(s) + Li_2NO_3 (aq)

    B. $Ca(NO_3)_2$ (aq) + Li_2SO_4 (aq) → $CaSO_4$ (s) + $2LiNO_3$ (aq)

    C. $Ca(NO_3)_2$ (aq) + Li_2SO_4 (aq) → $2LiNO_3$ (s) + $CaSO_4$ (aq)

    D. $Ca(NO_3)_2$(aq) + Li_2SO_4(aq) + $2H_2O$ (l) → $2LiNO_3$ (aq) + $Ca(OH)_2$ (aq) + H_2SO_4 (aq)

    E. None of the Above

43. Find the mass of CO_2 produced by combustion of 15 kg of isopropyl alcohol following the reaction:
    $2C_3H_7OH + 9O_2 → 6CO_2 + 8H_2O$

    A. 33 kg

    B. 44 kg

    C. 50 kg

    D. 60 kg

    E. 66 kg

44. What is the density of nitrogen gas at STP? Assume N_2 as an ideal gas and a value of 0.08206 L·atm/(mol·K) for the gas constant.

    A. 0.62 g/L

    B. 1.14 g/L

    C. 1.25 g/L

    D. 2.03 g/L

    E. 3.38 g/L

CHEMISTRY

45. Find the volume of methane that will produce 12 m³ of hydrogen in the reaction:

    $CH_4(g) + H_2O(g) \rightarrow CO(g) + 3H_2(g)$

    The temperature and pressure remain constant and assume ideal gases.

    A. 4.0 m³

    B. 32 m³

    C. 36 m³

    D. 64 m³

    E. Cannot be determined

46. Household "chlorine bleach" is sodium hypochlorite. Which of the following best represents the production of sodium hypochlorite, sodium chloride, and water by bubbling chlorine gas through aqueous sodium hydroxide?

    A. $4Cl(g) + 4NaOH(aq) \rightarrow NaClO_2(aq) + 3NaCl(aq) + 2H_2O(l)$

    B. $2Cl_2(g) + 4NaOH(aq) \rightarrow NaClO_2(aq) + 3NaCl(aq) + 2H_2O(l)$

    C. $2Cl(g) + 2NaOH(aq) \rightarrow NaClO(aq) + NaCl(aq) + H_2O(l)$

    D. $Cl_2(g) + 2NaOH(aq) \rightarrow NaClO(aq) + NaCl(aq) + H_2O(l)$

    E. None of the above

47. For the reaction
    $2M(s) + Cd^{2+} \rightarrow 2M^+ + Cd(s)$

    in an electrochemical cell at 25 °C. Which of the following are true?

    A. $M(s) \rightarrow M^+ + e^-$, takes place at the anode.

    B. As the reaction proceeds, the [M⁺] decreases.

    C. Cd^{2+} loses electrons.

    D. $Cd^{2+} + 2e^- \rightarrow Cd(s)$ takes place at the anode.

    E. There is not enough information to determine.

48. Write the equilibrium expression K_{eq} for the reaction:

    $CO_2(g) + H_2(g) \rightarrow CO(g) + H_2O(l)$

    A. $\dfrac{[CO][H_2O]}{[CO_2][H_2]^2}$

    B. $\dfrac{[CO_2][H_2]}{[CO][H_2O]}$

    C. $\dfrac{[CO][H_2O]}{[CO_2][H_2]}$

    D. $\dfrac{[CO]}{[CO_2][H_2]}$

    E. None of the above

CHEMISTRY

49. The exothermic reaction
    $2NO\ (g) + Br_2\ (g) \rightarrow 2NOBr\ (g)$ is at equilibrium. According to Le Chatelier's principle:

    A. Adding Br_2 will increase [NO].

    B. Adding [NO] will increase Br_2.

    C. An increase in container volume (with T constant) will increase [NOBr].

    D. An increase in pressure (with T constant) will increase [NOBr].

    E. An increase in temperature (with P constant) will increase [NOBr].

50. The equilibrium constant of the following reaction at a certain temperature is $K_{eq} = 2 \times 10^3$.
    $2NO\ (g) \rightarrow N_2\ (g) + O_2\ (g)$

    If a 1.0 L container at this temperature contains 90 mM N_2, 20 mM O_2, and 5 mM NO, what would occur?

    A. The reaction will produce more N_2 and O_2.

    B. The reaction is at equilibrium.

    C. The reaction will produce more NO.

    D. The temperature, T, is required to solve this problem.

    E. None of the above

51. $BaSO_4$ ($K_{sp} = 1 \times 10^{-10}$) is added to pure H_2O. How much is dissolved in 1 L of a saturated solution?

    A. 2 mg

    B. 10 ug

    C. 2 ug

    D. 100 pg

    E. Cannot be determined from the given information.

52. What are the pH and the pOH of 0.01 M HNO_3 (*aq*)?

    A. pH = 1.0, pOH = 9.0

    B. pH = 2.0, pOH = 12.0

    C. pH = 2.0, pOH = 8.0

    D. pH = 8.0, pOH = 6.0

    E. pH = 0.1, pOH = 0.9

CHEMISTRY

53. Which statements about reaction rates are true?

   I. Catalysts shift an equilibrium to favor formation of products
   II. Catalysts increase the rate of forward and reverse reactions
   III. A greater temperature increases the chance that a molecular collision will overcome a reaction's activation energy
   IV. A catalytic converter contains a homogeneous catalyst

   A. I and II

   B. II and III

   C. I, II, and III

   D. II, III, and IV

   E. I, III, and IV

54. Consider the reaction between iron and hydrogen chloride gas:
    $Fe(s) + 2HCl(g) \rightarrow FeCl_2(s) + H_2(g)$

    7 moles of iron and 10 moles of HCl react until the limiting reagent is consumed. Which statements are true?

    I. HCl is the excess reagent
    II. HCl is the limiting reagent
    III. 7 moles of H_2 are produced
    IV. 2 moles of the excess reagent remain

    A. I and III

    B. I and IV

    C. II and III

    D. II and IV

    E. I only

CHEMISTRY

55. The reaction:
$(CH_3)_3CBr(aq) + OH^-(aq) \rightarrow (CH_3)_3COH(aq) + Br^-(aq)$

occurs in three elementary steps:
$(CH_3)_3CBr \rightarrow (CH_3)_3C^+ + Br^-$ is slow

$(CH_3)_3C^+ + H_2O \rightarrow (CH_3)_3COH_2^+$ is fast

$(CH_3)_3COH_2^+ + OH^- \rightarrow (CH_3)_3COH + H_2O$ is fast

What is the rate law for this reaction?

A. Rate = $k\,[\,(CH_3)_3CBr\,]$

B. Rate = $k\,[OH^-]$

C. Rate = $k\,[(CH_3)_3CBr]\,[OH^-]$

D. Rate = $k\,[(CH_3)_3CBr]^2$

E. Rate = $\dfrac{k\,[(CH_3)_3CBr]}{[(CH_3)_3COH]}$

56. Which statement about thermochemistry is true?

A. Particles in a system move less freely at high entropy.

B. Water at 100 °C has the same internal energy as water vapor at 100 °C.

C. A decrease in the order of a system corresponds to an increase in entropy.

D. The Heat of Fusion is the energy needed to transform a liquid into a solid.

E. At its sublimation temperature, dry ice has higher entropy than CO_2 gas.

57. Which statement about reactions is true?

A. All spontaneous reactions are exothermic and cause an increase in entropy.

B. An endothermic reaction that increases the order of the system cannot be spontaneous.

C. A reaction can be non-spontaneous in both directions (forward and backward).

D. Melting snow is an exothermic process.

E. Thermodynamic functions are dependent on the reaction pathway.

CHEMISTRY

58. Given:

$E° = -2.37$ V for
$Mg^{2+} (aq) + 2e^- \rightarrow Mg (s)$

and

$E° = 0.80$ V for
$Ag^+ (aq) + e^- \rightarrow Ag (s)$

What is the standard potential of a voltaic cell composed of a piece of magnesium dipped in a 1 M Ag^+ solution and a piece of silver dipped in 1 M Mg^{2+} solution?

A. 0.77 V

B. 1.57 V

C. 3.17 V

D. 3.97 V

E. 0.38 V

59. What could cause this change in the energy diagram of a reaction (the energy scale is exactly the same for both figures)?

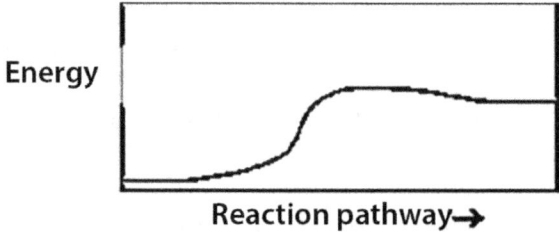

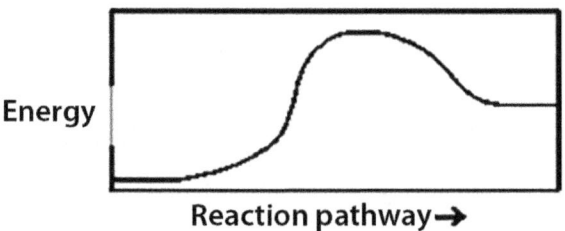

A. Adding a catalyst to an endothermic reaction

B. Removing a catalyst from an endothermic reaction.

C. Adding a catalyst to an exothermic reaction.

D. Removing a catalyst from an exothermic reaction.

E. Adding heat to an endothermic reaction.

CHEMISTRY

60. In the following phase diagram, _____ occurs as Pressure is decreased from A to B at constant Temperature and _____ occurs as Temperature is increased from C to D at constant Pressure.

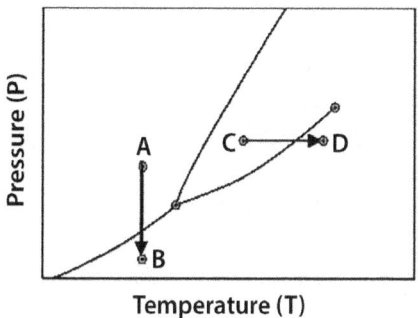

   A. deposition, melting
   B. sublimation, melting
   C. deposition, vaporization
   D. sublimation, vaporization
   E. melting, vaporization

61. Heat is added to a pure solid at its melting point until it all becomes liquid at its freezing point. Which of the following occur(s)?

   A. Intermolecular attractions are weakened.
   B. The kinetic energy of the molecules does not change.
   C. The freedom of the molecules in movement increases.
   D. The temperature of the system remains constant.
   E. All of the above

62. This compound contains an:

   A. alkene, carboxylic acid, ester, and ketone
   B. aldehyde, alkyne, ester, and ketone
   C. aldehyde, alkene, carboxylic acid, and ester
   D. acid anhydride, aldehyde, alkene, and amine
   E. aldehyde, amine, ketone, and alcohol

63. Which of the following pairs are isomers?

I.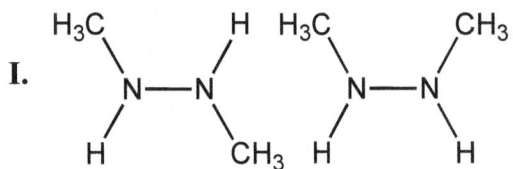

II. pentanal, 2-pentanone

III.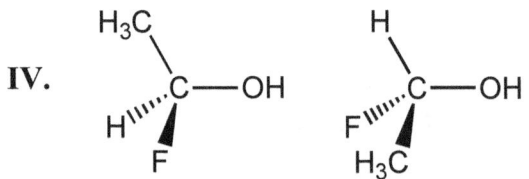

IV.
H3C\C—OH   H\C—OH
H/ F   F/ H3C

A. I and IV

B. II and III

C. I, II, and III

D. I, II, III, and IV

E. None of the above

64. Which of the following is **not** colligative property?

A. Viscosity lowering

B. Freezing point lowering

C. Boiling point elevation

D. Vapor pressure lowering

E. All of the above

65. A sample of 50.0 mL KOH is titrated with 0.100 M $HClO_4$. The initial buret reading is 1.6 mL and the reading at the endpoint is 22.4 mL. What is [KOH]?

A. 0.0416 M

B. 0.0481 M

C. 0.0832 M

D. 0.0962 mM

E. 0.0962 M

66. When KNO_3 dissolves in water, the water grows slightly colder. An increase in temperature will _____ the solubility of KNO_3.

A. increase

B. decrease

C. double

D. have no effect on

E. has an unknown effect with the information given on the solubility.

CHEMISTRY

67. Classify these biochemicals:

I. [structure of cytidine monophosphate]

II. [structure of a sugar]

III. [structure of a tripeptide]

IV. [structure of a triglyceride]

A. I - nucleotide, II - sugar, III - peptide, IV - fat

B. I - DNA, II - sugar, III - peptide, IV - lipid

C. I - disaccharide, II - sugar, III - fatty acid, IV - polypeptide

D. I - disaccharide, II - amino acid, III - fatty acid, IV - polysaccharide

E. I - nucleotide, II - sugar, III - triglyceride, IV – DNA

68. Which of the following can be determined from the Periodic Table?

I. The number of protons
II. The number of neutrons
III. The number of isotopes of that atom
IV. The number of valence electrons

A. I only

B. I and II

C. I, II, and III

D. I, II, III, and IV

E. I, II and IV only

69. Which of the following quantum numbers are needed to define the position of the electrons in an element?

A. principal, angular momentum, magnetic, and spin

B. principal, circular, magnetic and electromagnetic

C. angular, magnetic, electronic, spin

D. primary, angular momentum, magnetic, and spin

E. principal only

CHEMISTRY

70. Which of the following are true?

A. Anions are larger than their corresponding atom

B. Second Ionization energy is greater than the first Ionization Energy.

C. As you move down a group on the Periodic Table, the atomic radius increases.

D. Atoms with completed shells are more stable.

D. All of the above

71. The solubility of $CoCl_2$ is 54 g per 100 g of ethanol. Three flasks each contain 100 g of ethanol. Flask #1 also contains 40 g $CoCl_2$ in solution. Flask #2 contains 56 g $CoCl_2$ in solution. Flask #3 contains 5 g of solid $CoCl_2$ in equilibrium with 54 g $CoCl_2$ in solution. Which of the following describes the solutions present in the liquid phase of the flasks?

A. #1 - saturated, #2 - supersaturated, #3 - unsaturated

B. #1 - unsaturated, #2 - miscible, #3 - saturated

C. #1 - unsaturated, #2 - supersaturated, #3 - saturated

D. #1 - unsaturated, #2 - not at equilibrium, #3 - miscible.

E. #1 - unsaturated, #2 - saturated, #3 - miscible.

72. An experiment requires 100 mL of a 0.500 M solution of $MgBr_2$. How many grams of $MgBr_2$ will be present in this solution?

A. 9.21 g

B. 18.4 g

C. 11.7 g

D. 12.4 g

E. 15.6 g

73. Which of the following is most likely to dissolve in water?

A. H_2

B. CCl_4

C. SF_6

D. CH_3OH

E. CH_4

74. 10 kJ of heat are added to one kilogram of iron at 10 °C. What would be the final temperature? The specific heat of iron is 0.45 J/(g °C).

A. 22 °C

B. 27 °C

C. 32 °C

D. 37 °C

E. 14.5 °C

CHEMISTRY

75. A student wishes to prepare 4.0 liters of a 0.500 M KIO$_3$ (molar mass 214 g). The proper procedure is to weigh out:

   A 42.8 g of KIO$_3$, and add 4 kg of H$_2$O.

   B. 42.8 g of KIO$_3$ and add H$_2$O until the final homogenous solution has a volume of 4.0 L.

   C. 21.4 g of KIO$_3$ added to 4 L of water.

   D. 428 g of KIO$_3$ added to 4 L of water.

   E. 214 g of KIO$_3$ added to 4.0 L of H$_2$O

CHEMISTRY

ANSWER KEY

Question Number	Correct Answer	Your Answer	Question Number	Correct Answer	Your Answer	Question Number	Correct Answer	Your Answer
1	C		26	C		51	A	
2	B		27	A		52	B	
3	D		28	D		53	B	
4	A		29	C		54	D	
5	C		30	B		55	A	
6	D		31	A		56	C	
7	C		32	C		57	B	
8	B		33	E		58	C	
9	B		34	A		59	B	
10	C		35	C		60	D	
11	D		36	D		61	E	
12	A		37	A		62	C	
13	C		38	C		63	B	
14	D		39	D		64	A	
15	C		40	D		65	A	
16	B		41	C		66	A	
17	C		42	B		67	A	
18	A		43	A		68	E	
19	D		44	C		69	A	
20	D		45	A		70	E	
21	B		46	D		71	C	
22	C		47	A		72	A	
23	A		48	D		73	D	
24	A		49	D		74	C	
25	E		50	A		75	B	

CHEMISTRY

RATIONALES

1. A piston compresses a gas at constant temperature. Which gas properties increase?

   I. Average speed of molecules
   II. Pressure
   III. Molecular collisions with container walls per second

   A. I and II

   B. I and III

   C. II and III

   D. I, II, and III

   E. None of the above

The answer is C.
Assuming an ideal gas (PV = nRT) that could be in the molecular form like H_2 or atomic form like Ar. According to the equation, a decrease in volume (V) at constant temperature (T) will increase the pressure. With less volume (wall area) at higher pressure, collisions between the molecules or atoms will increase per second.

2. The temperature of a liquid is raised at atmospheric pressure. Which liquid property increases?

   A. Critical Pressure

   B. Vapor Pressure

   C. Surface Tension

   D. Viscosity

   E. Boiling Point

The answer is B.
The critical pressure of a liquid is its vapor pressure at the critical temperature and is always a constant value. If the temperature rises, the kinetic energy of molecules increases. This, in turn will decrease intermolecular attractions. More molecules will be free to escape to the vapor phase (vapor pressure increases), but the effect of attractions at the liquid-gas interface will fall (surface tension decreases) and molecules will flow against each other more easily (viscosity decreases).

CHEMISTRY

3. **Potassium crystallizes with two atoms contained in each unit cell. What is the mass of potassium found in a lattice 1.00×10^6 unit cells wide, 2.00×10^6 unit cells high, and 5.00×10^5 unit cells deep?**

   A. 85.0 μg

   B. 32.5 μg

   C. 64.9 μg

   D. 130 μg

   E. 130×10^6 μg

The answer is D.

The number of unit cells in a lattice is calculated by multiplying the number in each row, stack, and column:

1.00×10^6 unit cell lengths x $2.00\ 10^6$ unit cell lengths x 5.00×10^5 unit cell lengths = 1.00×10^{18} unit cells

Using Avogadro's number and the molecular weight of potassium (K), the mass could be calculated as follows:

$$1.00 \times 10^{18} \text{ unit cells} \times \frac{2 \text{ atoms of } K}{\text{unit cell}} \times \frac{1 \text{ mole of } K}{6.02 \times 10^{23} \text{ atoms of } K} \times \frac{39.098 g\ K}{1 \text{ mole of } K}$$

$$= 1.30 \times 10^{-4} g$$
$$= 130 \text{ μg}$$

CHEMISTRY

4. A gas is heated in a sealed container. Which of the following occur(s)?

A. Gas pressure rises

B. Gas density decreases

C. The average distance between molecules increases

D. The volume increases

E. All of the above

The answer is A.
The same material is kept in a constant volume, so neither density nor the distance between molecules will change. Pressure will rise due to the increase in molecular kinetic energy that in turn will impact container walls.

Mathematically, assuming an ideal gas (PV = nRT), if T increases at constant volume, pressure should increase because number of molecules, n, and gas constant, R, are constant values.

5. How many molecules are in 2.20 pg of a protein with a molecular weight of 150 kDa?

A. 8.83×10^9

B. 1.82×10^9

C. 8.83×10^6

D. 1.82×10^6

E. 8.83×10^{15}

The answer is C.
The prefix "p" for "pico-" indicates $\times 10^{-12}$ g. A kilo-Dalton is 1000 atomic mass units. To determine number of moles, mass has to be divided on molecular mass ($2.2 \times 10^{-12} / 150 \times 10^3 = 1.46 \times 10^{-17}$ moles). Then, the value has to be multiplied by Avogadro's number to determine number of molecules ($1.46 \times 10^{-17} \times 6.02 \times 10^{23} = 8.83 \times 10^6$).

CHEMISTRY

6. At STP, 20 μL of O_2 contain 5.4×10^{16} molecules. According to Avogadro's hypothesis, how many molecules are in 20 μL of Ne at STP?

   A. 5.4×10^{15}

   B. 1.0×10^{16}

   C. 2.7×10^{16}

   D. 5.4×10^{16}

   E. 1.3×10^{6}

The answer is D.
Avogadro's hypothesis states that equal volumes of different gases at the same temperature and pressure contain equal numbers of molecules.

Mathematically, using PV = nRT for both gases ($PV_1 = n_1RT$ and $PV_2 = n_2RT$, where P, R and T are same for both gases), and by injecting one equation into the other, it will possible to determine n_2.

7. An ideal gas at 50.0 °C and 3.00 atm is enclosed in a 300 cm³ cylinder. The cylinder volume changed by moving a piston until the gas reaches a T and P of 50.0 °C and 1.00 atm. What is the final volume of the gas?

   A. 100 cm³

   B. 450 cm³

   C. 900 cm³

   D. 1.20 dm³

   E. 150.0 cm³

The answer is C.
A three-fold decrease in pressure of a constant quantity of gas at constant temperature will cause a three-fold increase in gas volume.

Mathematically, you can always use PV = nRT to determine unknown parameters like volume in this question.

CHEMISTRY

8. **1-butanol, ethanol, methanol, and 1-propanol are all liquids at room temperature. Rank them in order of increasing viscosity.**

   A. 1-butanol < 1-propanol < ethanol < methanol

   B. methanol < ethanol < 1-propanol < 1-butanol

   C. methanol < ethanol < 1-butanol < 1-propanol

   D. 1-propanol < 1-butanol < ethanol < methanol

   E. Ethanol < methanol < 1 –butanol. 1-propanol

The answer is B.
Elevated viscosities result from strong intermolecular attractive forces. The listed molecules are all alcohols with -OH functional group attached to the end of a straight-chain alkane. In other words, they all have the formula $CH_3(CH_2)_{n-1}OH$ and the only difference between them is the length of the alkane corresponding to **n**. With all else identical, larger molecules have greater intermolecular attractive forces due to higher molecular surface for the attractions. Therefore the more **n** is greater, the more viscosity is higher. As a result: methanol (CH_3OH) < ethanol (CH_3CH_2OH) < 1-propanol ($CH_3CH_2CH_2OH$) < 1-butanol ($CH_3CH_2CH_2CH_2OH$).

CHEMISTRY

9. One mole of an ideal gas at STP occupies 22.4 L. At what temperature will one mole of an ideal gas at 1 atm occupy 31.0 L?

   A. 34.6 °C

   B. 105 °C

   C. 378 °C

   D. 442 °C

   E. 28 °C

The answer is B.
Either Charles' law, the combined gas law, or the ideal gas law could be used with temperature in Kelvin. Charles' law or the combined gas law with $P_1 = P_2$ may be manipulated to equate a ratio between temperature and volume when P and n are constants:

$$V \propto T \text{ or } \frac{P_1 V_1}{T_1} = \frac{P_2 V_2}{T_2} \Rightarrow \frac{T_1}{V_1} = \frac{T_2}{V_2} \Rightarrow T_2 = V_2 \frac{T_1}{V_1}$$

$$T_2 = 31.0 \text{ L} \frac{273.15 K}{22.4 L} = 378 K = 105 °C$$

The ideal gas law may also be used with the appropriate gas constant:

$$PV = nRT \Rightarrow T = \frac{PV}{nR}$$

$$T = \frac{(1\ atm)(31.0 L)}{(1\ mol)(0.08206 \frac{L-atm}{mol-K})} = 378 \text{ K} = 105 °C$$

CHEMISTRY

10. **What pressure is exerted by a mixture of 2.7 g of H₂ and 59 g of Xe at STP on a 50 L container?**

    A. 0.69 atm

    B. 0.76 atm

    C. 0.80 atm

    D. 0.97 atm

    E. 27.0 atm

The answer is C.
Grams of gas are first converted to moles (n = m/M):

$$2.7g\ H_2 \times \frac{1\ mol\ H_2}{2 \times 1.0079g\ H_2} = 1.33\ mol\ H_2 \text{ and } 59g\ Xe \times \frac{1\ mol\ H_2}{131.29g\ H_2} = 0.449\ mol\ Xe$$

Dalton's law of partial pressures for an ideal gas could be used to determine pressure of the mixture:

$$P_{total}V = (n_{H_2} + n_{Xe})RT \Rightarrow P_{total} = \frac{(n_{H_2} + n_{Xe})RT}{V}$$

$$P_{total} = \frac{(1.33\ mol + 0.449\ mol)\left(0.08206\ \frac{L-atm}{mol-K}\right)(273.15K)}{50L} = 0.80\ atm$$

11. **The normal boiling point of water on the Kelvin scale is closest to:**

    A. 112 °K

    B. 212 °K

    C. 273 °K

    D. 373 °K

    E. 298 °K

The answer is D.
Temperature in Kelvin is equal to the temperature in Celsius plus 273.15. Since the normal boiling point of water is 100 °C, hence it will boil at 373.15 °K (answer D is correct).

CHEMISTRY

12. Which phase may be present at the triple point of a substance?

    I. Gas
    II. Liquid
    III. Solid
    IV. Supercritical fluid

    A. I, II, and III

    B. I, II, and IV

    C. II, III, and IV

    D. I, II, III, and IV

    E. I, III, IV

The answer is A.
Gas, liquid, and solid may exist together at the triple point.

13. The solubility of pure CO_2 in water at 25 °C and 1.0 atm is 0.034 M. According to Henry's law, what is the solubility of pure CO_2 in water at 25 °C and 4.0 atm? Assume no chemical reaction occurs between CO_2 and H_2O.

    A. 0.0085 M

    B. 0.034 M

    C. 0.14 M

    D. 0.25 M

    E. 0.15 M

The answer is C.
Henry's law states that solubility of CO_2 in mol/L is proportional to partial pressure of the gas. A four-fold increase in pressure from 1.0 atm to 4.0 atm will increase the solubility by four-fold from 0.034 M to 0.14 M.

CHEMISTRY

14. **Which statement about molecular structures is false?**

    A. $H_2C=CH-CH=CH_2$ (with H's shown) is a conjugated molecule.

    B. A bonding σ orbital connects two atoms by a straight line between them.

    C. A bonding π orbital connects two atoms in a separate region from the straight line between them.

    D. $\left[\begin{array}{c} H \\ O=C-\ddot{O}: \end{array} \right]^- \longleftrightarrow \left[\begin{array}{c} H \\ :\ddot{O}-C=O: \end{array} \right]^-$

    The anion with resonance forms always exists in one form or the other.

    E. They are all false

The answer is D.
A conjugated molecule is a molecule with double bonds on adjacent atoms such as the molecule shown in A. Choices B and C give the definition of sigma and pi molecular orbitals. D is false because a resonance form resembles Lewis structures, but these structures do not describe the real state of the molecule, where the anion exists in a state between the two forms.

CHEMISTRY

15. What is the chemical composition of magnesium nitrate?

A. Mg (11.1%), N (22.2%), O (66.7%)

B. Mg (50.1%), N (22.2%), O (33.0%)

C. Mg (16.4%), N (18.9%), O (64.7%)

D. Mg (20.9%), N (24.1%), O (55.0%)

E. Mg (28.2%), N (16.2%), O (55.7%)

The answer is C.
First, the formula for magnesium nitrate is composed of magnesium and nitrate. Mg is an alkali earth metal and will always have a +2 charge (Mg^{2+}). The nitrate ion has -1 charge (NO_3^-). Hence, to balance the charge within the molecule, two nitrate ions are required for each Mg^{2+} ion. Therefore the correct formula of magnesium nitrate is $Mg(NO_3)_2$.

Second, after finding the correct chemical formula, it will easy to determine the chemical composition following the steps.

- Determine the number of atoms in $Mg(NO_3)_2$: 1Mg, 2N, 6O.
- Multiply by the molecular weight of the elements to determine the grams of each in one mole of the formula.

$$\frac{1 \text{ mol Mg}}{\text{mol Mg(NO}_3)_2} \times \frac{24.3 \text{ g Mg}}{\text{mol Mg}} = 24.3 \text{ g Mg/mol Mg(NO}_3)_2$$

$$2(14.0) = 28.0 \text{ g N/mol Mg(NO}_3)_2$$
$$6(16.0) = 96.0 \text{ g O/mol Mg(NO}_3)_2$$
$$\overline{148.3 \text{ g Mg(NO}_3)_2/\text{mol Mg(NO}_3)_2}$$

- Determine formula mass

- Divide to determine % composition

$$\%Mg = \frac{24.3 \text{ g Mg/mol Mg(NO}_3)_2}{148.3 \text{ g Mg(NO}_3)_2/\text{mol Mg(NO}_3)_2} = 0.164 \text{ g Mg/g Mg(NO}_3)_2 \times 100\% = 16.4\%$$

$$\%N = \frac{28.0}{148.3} \times 100\% = 18.9\% \qquad \%O = \frac{96.0}{148.3} \times 100\% = 64.7\%$$

Answer A represents the fractions of each atom in the formula and composition is based on mass percentage. Answer B does not use the correct formula for $Mg(NO_3)_2$. Answer D represents the chemical composition of $Mg(NO_2)_2$. Finally, answer E shows the chemical composition of $MgNO_3$, where it can be seen that the formula is charge unbalanced (Mg^{2+} requires $2NO_3^-$ not $1NO_3^-$).

CHEMISTRY

16. How many neutrons are there in $^{60}_{27}Co$?

A. 27

B. 33

C. 60

D. 87

E. 14

The answer is B.
The number of neutrons is calculated by subtracting the atomic number (27) from the mass number (60).

17. Select the list of atoms that is arranged in order of increasing size.

A. Mg, Na, Si, Cl

B. Si, Cl, Mg, Na

C. Cl, Si, Mg, Na

D. Na, Mg, Si, Cl

E. Mg, Si, Cl, Na

The answer is C.
These atoms are all in the same row of the Periodic Table. Size increases further to the left for atoms in the same row, which makes answer C correct.

CHEMISTRY

18. Based on trends in the Periodic Table, which of the following properties would you expect to be greater for Rb than for K?

    I. Density
    II. Melting point
    III. Ionization energy
    IV. Oxidation number in a compound with chlorine

    A. I only

    B. I, II, and III

    C. II and III

    D. I, II, III, and IV

    E. None of the above

The answer is A.

Rb is underneath K in the alkali metal column (group 1) of the Periodic Table. There is a general trend for density to increase lower on the table for elements in the same row, hence choice I is correct. Rb and K experience metallic bonds for intermolecular forces, and the strength of these bonds decrease for larger atoms further down the periodic table resulting in a lower melting point for Rb, thus II is incorrect. Ionization energy decreases for larger atoms further down the periodic table, hence III is also incorrect. Both Rb and K would be expected to have a charge of +1 and therefore an oxidation number of +1 in a compound with chlorine, so IV is incorrect.

CHEMISTRY

19. Why does CaCl₂ have a higher normal melting point than NH₃?

A. London dispersion forces in CaCl₂ are stronger than covalent bonds in NH₃.

B. Covalent bonds in NH₃ are stronger than dipole-dipole bonds in CaCl₂.

C. Ionic bonds in CaCl₂ are stronger than London dispersion forces in NH₃.

D. Ionic bonds in CaCl₂ are stronger than hydrogen bonds in NH₃.

E. None of the Above

The answer is D.
London dispersion forces are weaker than covalent bonds, which disregards choice A. A higher melting point will result from stronger intermolecular bonds that also eliminates choice B. CaCl₂ is a solid based on ionic bonds formed between Ca-Cl atoms, where the cation Na⁺ is located on the left and an anion Cl⁻ on the right of the Periodic Table. Since both Na and Cl are far from each other in the Periodic Table, this will induce strong ionic bonding. On the other hand, the dominant attractive forces between NH₃ molecules are mainly based on hydrogen bonds.

20. Rank the following bonds from least to most polar:

C-H, C-Cl, H-H, C-F

A. C-H < H-H < C-F < C-Cl

B. H-H < C-H < C-F < C-Cl

C. C-F < C-Cl < C-H < H-H

D. H-H < C-H < C-Cl < C-F

E. H-H < C-CL < C-H < C-F

The answer is D.
Bonds between atoms of the same element are purely non-polar. Hence H-H is the least polar bond in the list, which eliminates both choices A and C. Because of the small difference in electronegativity between C and H atoms, the C-H bond is considered as non-polar even though the electrons of the bond are slightly unequally shared. C-Cl and C-F are both polar covalent bonds due to the large difference in electronegativity of these elements, but C-F is more strongly polar because F has a greater electronegativity.

CHEMISTRY

21. In C₂H₂, each carbon atom contains the following valence orbitals:

A. *p* only

B. *p* and *sp* hybrids

C. *p* and *sp²* hybrids

D. *sp³* hybrids only

E. *s*, *p*, and *sp²* hybrids

The answer is B.
A carbon atom C has the following valence electronic configuration in the last layer: $2s^2 2p^2$. Before formation of a bond with another element like H or another C, one electron of the s orbital is promoted to the p orbital that is empty to yield the following configuration: $2s^1 2p^3$. Under this electronic configuration, C is ready to form 4 bonds. In C_2H_2, each C atom bonds to another C atom and an H. Bonding to two other atoms is achieved by combination into two p orbitals and two sp hybrids.

22. The boiling points of N₂, O₂ and Cl₂ are respectively -196, -182, and -34 °C. Cl₂ boils at a much higher temperature than expected. This is explained by:

A. Single bonds are easier to break than double or triple bonds.

B. Cl₂ has a longer bond length than the others.

C. Cl₂ has greater London Dispersion forces because it has more electrons.

D. Hydrogen bonding is stronger in Cl₂.

E. None of the above

The answer is C
Liquid oxygen, liquid nitrogen and liquid chlorine are all non-polar substances because they are composed of two similar atoms. Thus, they only experience London dispersion forces of attraction. These forces are greater for Cl_2 because it has more electrons, so has the highest boiling point.

CHEMISTRY

23. Which of the following statements regarding molecular geometries are true?

I. Double bonds are planar
II. A central atom in *sp²* hybridization is trigonal
III. A central atom with *sp²* hybridization of bipyramidal
IV. Single bonds are always planar

A. I and II

B. I and IV

C. II and IV

D. II and III

E. I, II, and IV

The answer is A.
Answer I is true because double bonds are planar and single bonds are not planar but linear. Answer II is true because sp² hybridization is trigonal. Therefore, only I and II are true.

24. The electronic configuration of Iron (Fe) is:

A. $1s^2\ 2s^2\ 2p^6\ 3s^2\ 3p^6\ 4s^2\ 3d^6$

B. $1s^2\ 2s^2\ 2p^6\ 3s^2\ 3p^6\ 3d^8$

C. $1s^2\ 2s^2\ 2p^6\ 3s^2\ 3p^6\ 4s^2$

D. $1s^2\ 2s^2\ 2p^6\ 3s^2\ 3p^6\ 4s^2\ 4p^6$

E. $1s^2\ 2s^2\ 2p^6\ 3s^2\ 3p^6\ 4s^2\ 4p^3\ 4d^3$

The answer is A.
Fe has 26 electrons. Filling each shell from 1*s* 2*s* 2p 3s 3p … then note that 4s is filled before 3d. The reason for this lies in nature of the element. Fe is a metal from the *d* block in the Periodic Table that has tendency to use electrons of *d* orbital to form bonds with other elements. Since energetically speaking, 3*d* is lower in energy than 4s, so Fe when reacting to lose electrons like in Fe^{2+} or Fe^{3+}, it is energetically favorable to lose them from the *d* orbital rather than from 4*s* with higher energy. Hence, this representation makes more sense when Fe is approached by other elements to form chemical bonds like in $FeCl_2$.

CHEMISTRY

25. Which of the following is a correct electron arrangement for oxygen?

A.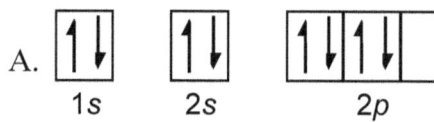

B. $1s^2 1p^2 2s^2 2p^2$

C. 2, 2, 4

D. 2, 2, 4, $\frac{1}{2}$

E. None of the above

The answer is E.
Choice A violates Hund's rule where all orbitals of *p* should be filled first with one electron that has positive spin before completing the orbital with secondary electron at opposite spin. In other words, the two electrons on the far right should occupy the final two orbitals. Choice B should have $1s^2 2s^2 2p^4$, as there is no 1*p* subshell. Choice C should be 2, 6, where the number indicate electrons in atom shells. Choice D is also wrong since there is no value ½, except for electron spin (+1/2, -1/2) which is another matter.

CHEMISTRY

26. Which of the following is a proper Lewis dot structure of CHClO?

A. [Lewis structure with O double-bonded to C, H and Cl single-bonded to C]

B. [Lewis structure with O double-bonded to C, H and Cl single-bonded to C, H with octet]

C. [Lewis structure with O double-bonded to C, H and Cl single-bonded to C]

D. [Lewis structure with O single-bonded to C, H single-bonded to C, Cl double-bonded to C]

E. None of the above

The answer is C.
Carbon C has a-4 valence shell electrons, H has 1, Cl has 7, and O has 6. As a result, the molecule has a total of 18 valence shell electrons, which eliminates choice B with 24. Choice B is also incorrect because it has an octet (8 electrons) around H instead of 2 electrons and there are only six electrons surrounding the central carbon and should be eight electrons on basis of 4 bonds with other elements. Choice A is incorrect as well since there are only 6 electrons surrounding the central C atom. A double bond between C and O gives the correct answer C. Finally, choice D is also incorrect because a total of 10 electrons surround the central atom C and Cl forms a double bond with C, which is incorrect because Cl has the capability to form one single bond and the rest of the electrons in shell are represented as unsharred electrons doublets (3 doublets in Cl).

CHEMISTRY

27. **Which intermolecular attraction force explains the following trend in straight-chain alkanes?**

Condensed structural formula	Boiling point (°C)
CH_4	-161.5
CH_3CH_3	-88.6
$CH_3CH_2CH_3$	-42.1
$CH_3CH_2CH_2CH_3$	-0.5
$CH_3CH_2CH_2CH_2CH_3$	36.0
$CH_3CH_2CH_2CH_2CH_2CH_3$	68.7

A. London dispersion forces

B. Hydrophobic interactions

C. Dipole-dipole interactions

D. Hydrogen bonding

E. Ion-induced dipole interactions

The answer is A.
Alkanes are composed entirely of non-polar C-C and C-H bonds with negligible polarity, resulting in no dipole interactions or hydrogen bonding (eliminating answers C and D). On the other hand, this list of alkanes do not form ionic bonds but only covalent bonding, which eliminates choice E. Hydrophobic interactions result from the tendency of nonpolar molecules like the list of these alkanes in a polar solvent like water to interact with one another and since no solvent is included in the list of the proposed alkanes, so we can also exclude answer B. Finally, London dispersion forces increase with the size of the molecule, resulting in a higher temperature requirement to break these bonds and a higher boiling point, hence the correct answer is A.

CHEMISTRY

28. Match the theory with the scientist who first proposed it:

   I. Electrons, atoms, and all objects with momentum also exist as waves.
   II. Electron density may be accurately described by a single mathematical equation.
   III. There is an inherent indeterminacy in the position and momentum of particles.
   IV. Radiant energy is transferred between particles in exact multiples of a discrete unit.

   A. I - de Broglie, II - Planck, III Schrödinger, IV - Thomson

   B. I - Dalton, II - Bohr, III - Planck, IV de Broglie

   C. I - Henry, II - Bohr, III - Heisenberg, IV - Schrödinger

   D. I - de Broglie, II - Schrödinger, III Heisenberg, IV - Planck

   E. I - Schrödinger, II - de Broglie, III - Plank, IV - Heisenberg

The answer is D.
Henry's law relates gas partial pressure to liquid solubility. Schrödinger's equation is the equation that describes electron density. Heisenberg developed the Uncertainty Principle and, finally Plank theorized discrete energy levels. Consequently, D is the correct answer.

29. The terrestrial composition of an element is: 50.7% as a stable isotope with an atomic mass of 78.9 μ and 49.3% as a stable isotope with an atomic mass of 80.9 μ. Calculate the atomic mass of the element.

   A. 79.0 μ

   B. 79.8 μ

   C. 79.9 μ

   D. 80.8 μ

   E. 80.0 μ

The answer is C.
Atomic mass of element = (Fraction as 1st isotope) (Atomic mass of 1st isotope)
$$+$$
(Fraction as 2nd isotope) (Atomic mass of 2nd isotope)
$$= (0.507)(78.9\ u) + (0.493)(80.9\ u)$$
$$= 79.89\ u$$
$$= 79.9\ u$$

CHEMISTRY

30. $^{3}_{1}H$ decays with a half-life of 12 years. 3.0 g of pure $^{3}_{1}H$ were placed in a sealed container 24 years ago. How many grams of $^{3}_{1}H$ remain?

A. 0.38 g

B. 0.75 g

C. 1.5 g

D. 0.125 g

E. 3.0 g

The answer is B.
Every 12 years (half-time), the amount is cut into half, yielding a remaining mass of (3g/2 = 1.5 g). After an additional 12 years (total of 24 years), 0.75 g will remain in the container (1.5/2 = 0.75 g).

31. Moving down a column on the Periodic Table:

I. The atomic radius increases
II. Ionization energy increases
III. Protons are added
IV. Metallic characteristics increase

A. I, II, III

B. I and II only

C III only

D. III and IV only

E. None of the above

The answer is A.
Moving down a column in the Periodic Table increases the atomic radii, ionization energy, and protons are added. On the other hand, the metallic characteristics do not increase.

CHEMISTRY

32. Moving from left to right on a Periodic Table (Li to Ne) which of the following statements are true:

    I.   The atomic radius decreases
    II. Electrons are added
    III. Ionization energies increases
    IV. Electronegativity decreases

    A. I only

    B. I and II

    C. I, II, and III

    D. I, II, III, and IV

    E. None of the above

The answer is C.
Moving from left to right in the Periodic Table decreases the atomic radius and electrons are added. On the other hand, ionization energies and electronegativities increase.

CHEMISTRY

33. NH₄F is dissolved in water. Which of the following are the conjugate acid/base pairs present in the solution?

   I. NH_4^+/NH_4OH
   II. HF/F^-
   III. H_3O^+/H_2O
   IV. H_2O/OH^-

   A. I only

   B I, II, and III

   C. I, III, and IV

   D. II and IV

   E. II, III, and IV

The answer is E.
NH₄F is soluble in water and fully dissociates to form NH_4^+ and F^-. F^- is a weak base with HF as its conjugate acid (II). NH_4^+ is a weak acid with NH_3 as its conjugate base. A conjugate acid/base pair must have the form HX/X, where the proton is transferred between HX and X. NH_4^+/NH_4OH (I) is not a conjugate acid/base pair because no proton is transferred between NH_4^+ and NH_4OH, but OH^- was added to NH_4^+. This eliminates choices A and B. H_3O^+/H_2O and H_2O/OH^- (III and IV) are always present in water and in all aqueous solutions as conjugate acid/base pairs, where both of them undergo transfer of a proton. The following equilibrium reactions occur in $NH_4F_{(aq)}$:

$NH_4^+{}_{(aq)} + OH^-{}_{(aq)} = NH_{3(aq)} + H_2O_{(l)}$
$F^-{}_{(aq)} + H_3O^+{}_{(aq)} = HF_{(aq)} + H_2O_{(l)}$
$2H_2O_{(l)} = H_3O^+{}_{(aq)} + OH^-{}_{(aq)}$

CHEMISTRY

34. Rank the following from lowest to highest pH. Assume a small volume for the component given in moles:

    I.   0.01 mol HCl added to 1 L H_2O
    II.   0.01 mol HI added to 1 L of an acetic acid/sodium acetate solution at pH 4.0
    III. 0.01 mol NH_3 added to 1 L H_2O
    IV. 0.1 mol HNO_3 added to 1 L of a 0.1 M $Ca(OH)_2$ solution

A. I < II < III < IV

B. I < II < IV < III

C. II < I < III < IV

D. II < I < IV < III

E. IV < III < II < I

The answer is A.
Solution I: HCl is a strong acid. Therefore Solution I has a pH of 2 because:
$$pH = -\log_{10}[H^+] = -\log_{10}(0.01) = 2$$
Solution II: HI is also a strong acid and would have a pH of 2 at this concentration in water, but the buffer capacity made exercised by acetic acid/sodium acetate will prevent the pH from dropping this low. As a result, solution II will have a pH above 2 and below 4, eliminating choices C and D.

Solution III: If a strong base was in Solution III, its pOH would be 2. Using the equation pH + pOH = 14, results in pH equals to 12. Because NH_3 is a weak base, the pH of Solution III will be greater than 7 and less than 12.

Solution IV: A neutralization reaction occurs in Solution IV between 0.1 mol of H^+ from the strong acid HNO_3 and 0.2 mol of OH^- from the strong base $Ca(OH)_2$. Each mole of $Ca(OH)_2$ frees two equivalent moles of a base in the neutralization reaction. The base is the excess reagent; hence 0.1 mol of OH^- remains in the solution after the reaction is completed. This solution will have a pOH of 1 and a pH of 13.

As a result, the correct ranking follows choice A: pH 2 < pH (between 2 and 4) < pH (between 7 and 12) < pH 13.

CHEMISTRY

35. Which statement about acids and bases is **not** true?

A. All strong acids ionize in water

B. All Lewis acids accept an electron pair

C. All Brønsted bases use OH⁻ as a proton acceptor

D. All Arrhenius acids form H⁺ ions in water

E. Water can act as either an acid or a base

The answer is C.
Choice A corresponds to the definition of a strong acid. Choice B corresponds to the definition of a Lewis acid, and choice D is the definition of an Arrhenius acid. Water has an amphoteric character, meaning that it could play the role of an acid in alkaline medium and a base is an acidic medium. By definition, all Arrhenius bases form OH⁻ ions in water, and all Brønsted bases are proton acceptors. But not all Brønsted bases use OH⁻ as a proton acceptor. For example, NH_3 is a Brønsted base.

CHEMISTRY

36. What is the pH of a buffer solution made of 0.128 M sodium formate (HCOONa) and 0.072 M formic acid (HCOOH)? The pKa of formic acid is 3.75.

A. 2.0

B. 3.0

C. 3.75

D. 4.0

E. 5.0

The answer is D.
HCOOH is a weak acid and HCOO⁻ that results from dissociation of HCOONa, is its conjugate base.

From pKa, K_a of formic acid is calculated as: $K_a = 10^{-pK_a} = 10^{-3.75} = 1.78 \times 10^{-4}$

This corresponds to the equilibrium constant: $K_a = \frac{[H^+][HCOO^-]}{[HCOOH]} = 1.78 \times 10^{-4}$

For the dissociation reaction of formic acid: HCOOH → H⁺ + HCOO⁻
The pH could be estimated from calculating the concentration of H⁺ as follows:

$$[H^+] = K_a \frac{[HCOOH]}{[HCOO^-]} = (1.78 \times 10^{-4}) \frac{0.072}{0.128} = 1.0 \times 10^{-4} \text{ M}$$

$$pH = -log_{10}[H^+] = -log_{10}(1.0 \times 10^{-4}) = 4.0$$

CHEMISTRY

37. A 100 L vessel of pure O_2 at 500 kPa and 20 °C is used for the combustion of butane following the reaction: $2C_4H_{10} + 13O_2 \rightarrow 8CO_2 + 10H_2O$

    Find the mass of butane that should consume all the O_2 in the vessel. Assume O_2 is an ideal gas and use a value of $R = 8.314$ J/(mol•K).

    A. 183 g

    B. 467 g

    C. 1.83 kg

    D. 2.6 kg

    E. 7.75 kg

The answer is A.

The volume of O_2 is given and asked to determine the mass. The best way to proceed is to link the volume to moles which in turn be linked to mass. Since the gas is considered as ideal, hence PV = nRT is applicable. But first several units must be converted:

> Units of joules are identical to $m^3 \cdot Pa$
> 500 kPa is 500×10^3 Pa
> 100 L is 0.100 m^3
> 20 °C is 293.15 K

PV = nRT is rearranged to give:

$$n = \frac{PV}{RT} = \frac{(500 \times 10^3 \ Pa)(0.100 \ m^3 O_2)}{(8.314 \frac{m^3 \cdot Pa}{mol \cdot K})(293.15 \ K)} = 20.51 \ mol \ O_2$$

Conversion of moles into mass relies on stoichiometry. The molecular weight of butane is 58.1 u:

$$20.51 \ mol \ O_2 \times \frac{2 \ mol \ C_4H_{10}}{13 \ mol \ O_2} \times \frac{58.1 g \ C_4H_{10}}{1 \ mol \ C_4H_{10}} = 183 g \ C_4H_{10}$$

CHEMISTRY

38. **32.0 g of hydrogen and 32.0 grams of oxygen react together to form water until the limiting reagent is consumed. What reagents and products are present in the vessel after the reaction is complete?**

A. 16.0 g O_2 and 48.0 g H_2O

B. 24.0 g H_2 and 40.0 g H_2O

C. 28.0 g H_2 and 36.0 g H_2O

D. 28.0 g H_2 and 34.0 g H_2O

E. 28.0 g H_2 and 16.0 g O_2

The answer is C.
First, the reaction should be written and properly balanced: $2H_2 + O_2 \rightarrow 2H_2O$

A quick way of solving the problem is to think accordingly:
- One mole of H_2 is about 2.0 g, hence about 16 moles of H_2 are present.
- One mole of O_2 corresponds to 32.0 g, thus one mole of O_2 is present.
- If 16 moles of H_2 react with one mole of O_2, 2 moles of H_2 will be consumed before one mole of O_2. Hence, O_2 is the limiting reagent, which eliminates choice A.
- 16 moles less 2 moles results in 14 moles of H_2 or about 28 g, which eliminates choice B.
- The reaction started with a total of 64.0 g. Law of conservation of mass for chemical reactions forces the total final mass to be constant 64.0 g, which eliminates choice D.

The problem could also be solved using the standard method.

- First, mass is converted to moles:

$$32.0\text{g } H_2 \times \frac{1 \text{ mol } H_2}{2.016 g\, H_2} = 15.87 \text{ mol } H_2 \text{ and } 32.0\text{g } O_2 \times \frac{1 \text{ mol } O_2}{32.00 g\, O_2} = 1.000 \text{ mol } O_2$$

- Dividing by the stoichiometric coefficients gives:

$$15.87 \text{ mol } H_2 \times \frac{1 \text{ mol reaction}}{2 \text{ mol } H_2} = 7.935 \text{ mol reaction if } H_2 \text{ is limiting}$$

$$1.000 \text{ mol } O_2 \times \frac{1 \text{ mol reaction}}{2 \text{ mol } O_2} = 1.000 \text{ mol reaction if } O_2 \text{ is limiting}$$

- O_2 is the limiting reagent, thus O_2 will not remain in the vessel.

$$1.000 \text{ mol } O_2 \text{ consumed} \times \frac{2 \text{ mol } H_2O \text{ produced}}{1 \text{ mol } O_2} \times \frac{18.016 g\, H_2O}{1 \text{ mol } H_2O} = 36.0\text{g } H_2O \text{ produced}$$

$$1.000 \text{ mol } O_2 \text{ consumed} \times \frac{2 \text{ mol } H_2 \text{ consumed}}{1 \text{ mol } O_2} \times \frac{2.016 g\, H_2}{1 \text{ mol } H_2} = 4.03\text{g } H_2 \text{ consumed}$$

- The remaining H_2 could be estimated as follows:
32.0 g H_2 (initial) – 4.03 g H_2 (consumed) = 28.0 g H_2 (remaining)

CHEMISTRY

39. Which reaction is not a redox process?

A. Combustion of octane: $2C_8H_{18} + 25O_2 \rightarrow 16CO_2 + 18H_2O$

B. Depletion of a lithium battery: $Li + MnO_2 \rightarrow LiMnO_2$

C. Corrosion of aluminum by acid: $2Al + 6HCl \rightarrow 2AlCl_3 + 3H_2$

D. Taking an antacid for heartburn: $CaCO_3 + 2HCl \rightarrow CaCl_2 + H_2CO_3 \rightarrow CaCl_2 + CO_2 + H_2O$

E. None of the above

The answer is D.
The oxidation state of atoms is altered in a redox process. During combustion (choice A), the carbon atoms are oxidized from oxidation number of -4 to +4. Oxygen atoms are reduced from an oxidation number of 0 to -2. All batteries (choice B) generate electricity by forcing electrons involved in the redox processes to flow through an external circuit. Li is oxidized from 0 in the metal to +1 in the $LiMnO_2$ salt. Mn is reduced from +4 in manganese (IV) oxide to +3 in lithium manganese (III) oxide salt. Corrosion (choice C) is due to oxidation with an oxidant like O_2. Al is oxidized from 0 to +3. H is reduced from +1 to 0. Acid-base neutralization (choice D) transfers a proton (an H atom with an oxidation state of +1) from an acid to a base. The oxidation state of all atoms remains unchanged (Ca at +2, C at +4, O at -2, H at +1, and Cl at -1), hence D is the correct answer. Note that choices C and D both involve an acid HCl. The availability of electrons in aluminum metal favors electron transfer but the availability of CO_3^{2-} as a proton acceptor favors proton transfer.

CHEMISTRY

40. Which of the following are true about Galvanic cells?

    I.   Two half reactions take place in separate two chambers
    II. Oxidation takes place at the anode
    III. Reduction takes place at the cathode
    IV. If the salt bridge is removed, the voltage drops to zero
    V.   Le Chatelier's principle can be applied to the systematic

    A. I, II, and III

    B. only II and III

    C. only I and IV

    D. All of the above

    E. None of the above

The answer is D.
All of the statements are true when learning about Galvanic cells.

CHEMISTRY

41. Balance the equation of the neutralization reaction between phosphoric acid and calcium hydroxide by filling in the blank stoichiometric coefficients from left to right.

    A. 4, 3, 1, 4

    B. 2, 3, 1, 8

    C. 2, 3, 1, 6

    D. 2, 1, 1, 2

    E. 4, 3, 1, 1

The answer is C.
The given equation is unbalanced, and the best way to proceed is determine the number of atoms on each side of the reaction then multiply by an appropriate number to balance it. For reactants (left side of the arrow): 5H, 1P, 6O, and 1Ca. For products (right side of the arrow): 2H, 2P, 9O, and 3Ca.

Assuming that the molecule with most atoms $Ca_3(PO_4)_2$ has a coefficient of one, the required coefficients to have the same number of atoms on each side of the equation could be guessed. Assuming $Ca_3(PO_4)_2$ has a coefficient of one, meaning that there will be 3Ca and 2P on the right side of the reaction because H_2O has no Ca or P. A balanced reaction would also have 3Ca and 2P on the left side of the reaction. This should be achieved by multiplying H_3PO_4 by 2 and $Ca(OH)_2$ by 3. This leads to the following:

$$2H_3PO_4 + 3Ca(OH)_2 \rightarrow Ca_3(PO_4)_2 + ?H_2O$$

The coefficient for H_2O is found by balancing H or O. Whichever one is chosen, the other atom should be checked to confirm that the reaction is well balanced. For H, there are 6H from $2H_3PO_4$ and 6 from $3Ca(OH)_2$ for a total of 12H on the left side. In turn, there must be 12H on the right side of the reaction to be balanced properly. None of these account for $Ca_3(PO_4)_2$, consequently all 12H must be associated with H_2O, which should have a coefficient of 6:

$$2H_3PO_4 + 3Ca(OH)_2 \rightarrow Ca_3(PO_4)_2 + 6H_2O$$

At the end, it is always useful to have a final look on the reaction if it is properly balanced. There are 8O from $2H_3PO_4$ and 6 from $3Ca(OH)_2$ for a total of 14 on the left side, and 8O from $Ca_3(PO_4)_2$ and 6 from $6H_2O$ for a total of 14 on the right side, consequently the equation is properly balanced.

An alternative method to balance the reaction would be to use the coefficients given for answers A, B, C, D, and E. This should be the quicker way to carry on under these circumstances since the answers are already provided.

CHEMISTRY

42. **Which of the following show the reaction between calcium nitrate and lithium sulfate in aqueous solution? Include all reagents and products, and make sure the reaction is mass balanced.**

    A. $CaNO_3(aq) + Li_2SO_4(aq) \rightarrow CaSO_4(s) + Li_2NO_3(aq)$

    B. $Ca(NO_3)_2(aq) + Li_2SO_4(aq) \rightarrow CaSO_4(s) + 2LiNO_3(aq)$

    C. $Ca(NO_3)_2(aq) + Li_2SO_4(aq) \rightarrow 2LiNO_3(s) + CaSO_4(aq)$

    D. $Ca(NO_3)_2(aq) + Li_2SO_4(aq) + 2H_2O(l) \rightarrow 2LiNO_3(aq) + Ca(OH)_2(aq) + H_2SO_4(aq)$

    E. None of the Above

The answer is B.
When two ionic compounds are in solution, a precipitation or complexation reaction should be considered. From the given names, it could be determined that the two reactants are the ionic compounds $Ca(NO_3)_2$ and Li_2SO_4. When dissolved in aqueous solution, the following ions are present in the solution: Ca^{2+}, NO_3^-, Li^+, and SO_4^{2-}, because the salts fully dissociates. Solubility rules indicate that nitrates are always soluble but sulfate forms solid precipitates with Ca^{2+} in the present case $CaSO_4$ (s). Choice A is incorrect because it assumes that the nitrate anion NO_3^- has a (2–) charge instead of its (1–) charge. Choice C assumes that lithium nitrate is the precipitate, so is also incorrect. Choice D assumes that mixture of both salts with water will form a sulfuric acid and calcium hydroxide, strong acid and base, in addition to lithium nitrates. However, if this reaction is feasible, H_2SO_4 should neutralize $Ca(OH)_2$ to form water again, hence D is also incorrect. As a result, B is correct and feasible reaction.

CHEMISTRY

43. Find the mass of CO_2 produced by combustion of 15 kg of isopropyl alcohol following the reaction:

$$2C_3H_7OH + 9O_2 \rightarrow 6CO_2 + 8H_2O$$

A. 33 kg

B. 44 kg

C. 50 kg

D. 60 kg

E. 66 kg

The answer is A.
It is always helpful to remember the rule "grams to moles to moles to grams".
- Step 1: Convert mass to moles for the known compound, which is C_3H_7OH (n = m/M = 15/60.1 = 0.249 moles).
- Step 2: Relate moles of known value to moles of the unknown value by their stoichiometry coefficients. Since the stoichiometry of the reaction indicates that 1 moles of is C_3H_7OH produces 3 moles of CO_2, hence total number of moles of CO_2 produced during the reaction are (0.249x3 = 0.747 moles).
- Step 3: Converting moles of the unknown value to a mass.
    - m = nM = 0.747x44= 33 gr

CHEMISTRY

44. What is the density of nitrogen gas at STP? Assume N₂ as an ideal gas and a value of 0.08206 L·atm/(mol·K) for the gas constant.

A. 0.62 g/L

B. 1.14 g/L

C. 1.25 g/L

D. 2.03 g/L

E. 3.38 g/L

The answer is C.
The molecular mass M of N_2 is 28.0 g/mol and since the gas is considered as ideal, thus PV = nRT. By rearranging and replacing density by d=m/V, the following equation could be reached:

$$d = \frac{nM}{V} = \frac{PM}{RT} = \frac{(1\ atm)(28.0\ \frac{g}{mol})}{(0.08206\ \frac{L.atm}{mol.K})(273.15\ K)} = 1.25\ \frac{g}{L}$$

On the other hand, choice A results if nitrogen is considered as an atomic gas N instead of molecular N_2 gas. Choice B is obtained by using a value of 25 °C for standard temperature, which is wrong because the temperature should be converted to Kelvin in STP.

Another method to solve the problem, which is relatively more quicker than the previous one is consider that one mole of an ideal gas at STP occupies 22.4 L, and by taking density for 1 mole (d = m/V = M/V), the following equation could be reached:

$$d\ (in\ \frac{g}{L}) = \frac{M\ (in\ \frac{g}{mol})}{22.4\ \frac{L}{mol}} = \frac{28.0\ \frac{g}{mol}}{22.4\ \frac{L}{mol}} = 1.25\ \frac{g}{L}$$

CHEMISTRY

45. Find the volume of methane that will produce 12 m³ of hydrogen in the reaction:

$$CH_4(g) + H_2O(g) \rightarrow CO(g) + 3H_2(g)$$

The temperature and pressure remain constant and assume ideal gases.

A. 4.0 m³

B. 32 m³

C. 36 m³

D. 64 m³

E. Cannot be determined

The answer is A.
The reaction is properly mass balanced; hence stoichiometric coefficients are used to determined number of moles of reactants and product involved in the chemical process. It could be seen that 1 mole methane produces 3 moles hydrogen. Because the gases are considered ideal, hence PV = nRT for both gases, where T, P, R are all constants:

$(PV = nRT)_{CH_4} \Rightarrow (V/n)_{CH_4} = RT/P$
$(PV = nRT)_{H_2} \Rightarrow (V/n)_{H_2} = RT/P$

Replacing RT/P in one of the equations gives: $(V/n)_{CH_4} = (V/n)_{H_2}$
Since 1 mole methane gives 3 moles hydrogen and final volume of hydrogen is 12 m³, thus the volume of methane equals to 4 m³.

CHEMISTRY

46. Household "chlorine bleach" is sodium hypochlorite. Which of the following best represents the production of sodium hypochlorite, sodium chloride, and water by bubbling chlorine gas through aqueous sodium hydroxide?

A. $4Cl(g) + 4NaOH(aq) \rightarrow NaClO_2(aq) + 3NaCl(aq) + 2H_2O(l)$

B. $2Cl_2(g) + 4NaOH(aq) \rightarrow NaClO_2(aq) + 3NaCl(aq) + 2H_2O(l)$

C. $2Cl(g) + 2NaOH(aq) \rightarrow NaClO(aq) + NaCl(aq) + H_2O(l)$

D. $Cl_2(g) + 2NaOH(aq) \rightarrow NaClO(aq) + NaCl(aq) + H_2O(l)$

E None of the above.

The answer is D.
Chlorine gas is a diatomic molecule Cl_2, which eliminates choices A and C. ClO_2^- is called chlorine dioxide, and the t hypochlorite ion is ClO^-. This eliminates choices A and B. Finally, it could be noticed that all of the equations are properly balanced.

47. For the reaction $2M(s) + Cd^{2+} \rightarrow 2M^+ + Cd(s)$ in an electrochemical cell at 25 °C. Which of the following are true?

A. $M(s) \rightarrow M^+ + e-$, takes place at the anode.

B. As the reaction proceeds, the $[M^+]$ decreases.

C. Cd^{2+} loses electrons.

D. $Cd^{2+} + 2e- \rightarrow Cd(s)$ takes place at the anode.

E. There is not enough information to determine.

The answer is A.
Choice A is true because $M(s)$ loses electrons; therefore this is an oxidation reaction that occurs at the anode. As the reaction proceeds the $[M^+]$ increase because it is a product, hence choice B is false. Cd^{2+} gains the electrons and the reduction reaction takes place at the cathode; therefore C and D are also false.

CHEMISTRY

48. Write the equilibrium expression K_{eq} for the reaction:

$$CO_2 (g) + H_2 (g) \rightarrow CO (g) + H_2O (l)$$

A. $\dfrac{[CO][H_2O]}{[CO_2][H_2]^2}$

B. $\dfrac{[CO_2][H_2]}{[CO][H_2O]}$

C. $\dfrac{[CO][H_2O]}{[CO_2][H_2]}$

D. $\dfrac{[CO]}{[CO_2][H_2]}$

E. None of the above

The answer is D.
By convention, a chemical reaction is always assumed to proceed from left to right, where left side are reactants and right side are products. In the present reaction, CO_2 (g) and H_2 (g) are the reactants and CO (g) and H_2O (l) are the products.

An equilibrium constant of a reaction is determined by multiplying the product concentrations together in the numerator and reactant concentrations in the denominator. This eliminates choice B because the notion of reactants and products are reversed. The reaction is mass balanced and indicates that the stoichiometric coefficient of H_2 is (1), which eliminates choice A. For heterogeneous reactions, concentrations of pure liquids or solids such as solvents that are in excess during the reaction are always put equals to unity (1), hence could be disregarded from the expression of the equilibrium constant because they have no influence on it (eliminates choice C). As a result, D the most correct answer.

CHEMISTRY

49. The exothermic reaction 2NO (*g*) + Br$_2$ (*g*) → 2NOBr (*g*) is at equilibrium. According to Le Chatelier's principle:

    A. Adding Br$_2$ will increase [NO].

    B. Adding [NO] will increase Br$_2$.

    C. An increase in container volume (with T constant) will increase [NOBr].

    D. An increase in pressure (with T constant) will increase [NOBr].

    E. An increase in temperature (with P constant) will increase [NOBr].

The answer is D.
Le Chatelier's principle predicts that a chemical reaction at equilibrium will shift to partially offset any change. In other words, if Br$_2$ is added to the reaction that is at equilibrium, this will partially offset by reducing [Br$_2$] and [NO] via a shift to the right side of the reaction, which eliminates choice A. If [NO] was added to the reaction, it will have same effect by shifting the reaction to the left side to produce more NOBr(g), hence choice B is incorrect. For the remaining possibilities, because the reaction is exothermic (generates heat), thus it could be simplified as: 3 moles = 2 moles + heat.

An increase in container volume will decrease pressure because changes in pressure are attributed to changes in volume. Because the reactant side has greater number of moles than does the product side, this change will be partially offset by an increase in number of moles present, shifting the reaction to the left side, which eliminates choice C. An increase in pressure will be offset by a decrease in number of moles present, shifting the reaction to the right side, making choice D the correct answer. Raising the temperature by adding heat will shift the reaction to the left side, making choice E incorrect.

CHEMISTRY

50. The equilibrium constant of the following reaction at a certain temperature is $K_{eq} = 2 \times 10^3$.

$$2NO\ (g) \rightarrow N_2\ (g) + O_2\ (g)$$

If a 1.0 L container at this temperature contains 90 mM N_2, 20 mM O_2, and 5 mM NO, what would occur?

A. The reaction will produce more N_2 and O_2.

B. The reaction is at equilibrium.

C. The reaction will produce more NO.

D. The temperature, T, is required to solve this problem.

E. None of the above

The answer is A.

The reaction quotient under these conditions could be calculated as:

$$Q = \frac{[N_2][O_2]}{[NO]^2} = \frac{(0.090\ M)(0.020\ M)}{(0.005\ M)^2} = 72$$

This value is lower than K_{eq} ($72 < 2 \times 10^3$), therefore $Q < K_{eq}$, meaning that the reaction did not yet reach the equilibrium, which eliminates choice B. To reach equilibrium, the numerator of Q must be larger than denominator and this would occur if more products are produced, hence eliminates choice C as well. Therefore, more NO will react to make more N_2 and O_2, and choice A is correct. There is no need to know temperature to solve this problem since K_{eq} is already provided.

CHEMISTRY

51. BaSO$_4$ (K_{sp} = 1x10^{-10}) is added to pure H$_2$O. How much is dissolved in 1 L of a saturated solution?

   A. 2 mg

   B. 10 ug

   C. 2 ug

   D. 100 pg

   E. Cannot be determined from the given information.

The answer is A.
The solubility reaction of BaSO$_4$ (s) is written as follows:
BaSO$_4$ (s) → Ba^{2+} (aq) + SO$_4^{2-}$ (aq)

At saturation, K$_{sp}$ = [Ba^{2+}][SO$_4^{2-}$]

Therefore, in a saturated solution, the concentration of species could be estimated from the solubility constant: $[Ba^{2+}] = [SO_4^{2-}] = \sqrt{1 \times 10^{-10}}$ = 1 x 10^{-5} M
The mass in one liter is found from molarity (c = m/MV ⇒ m = cMV):

$$1 \times 10^{-5} \frac{mol\ Ba^{2+}\ or\ SO_4^{2-}}{L} \times \frac{1\ mol\ dissolved\ BaSO_4}{1\ mol\ Ba^{2+} or\ SO_4^{2-}} \times \frac{(137+32+4 \times 16)g\ BaSO_4}{1\ mol\ BaSO_4}$$

$$= 0.002 \frac{g}{L} BaSO_4 \times 1L\ solution \times \frac{1000\ mg}{g} = 2mg\ BaSO_4$$

CHEMISTRY

52. What are the pH and the pOH of 0.01 M HNO₃ (aq)?

A. pH = 1.0, pOH = 9.0

B. pH = 2.0, pOH = 12.0

C. pH = 2.0, pOH = 8.0

D. pH = 8.0, pOH = 6.0

E. pH = 0.1, pOH = 0.9

The answer is B.
HNO_3 is a strong acid, thus fully dissociates in solution to form H^+ and NO_3^- following the reaction: $HNO_3 \rightarrow H^+ + NO_3^-$. Since the stoichiometry of the reaction indicates 1 moles of HNO_3 will form 1 mole H^+ and 1 mole NO_3^-, hence:
$$[H^+] = 0.010M = 1.0 \times 10^{-2}M$$
$$pH = -\log_{10}[H^+] = -\log_{10}(1.0 \times 10^{-2}) = 2.0 \text{ (choices B or C)}$$
From pH + pOH = 14: pOH = 12.0 (choice B)

53. Which statements about reaction rates are true?

I. Catalysts shift an equilibrium to favor formation of products
II. Catalysts increase the rate of forward and reverse reactions
III. A greater temperature increases the chance that a molecular collision will overcome a reaction's activation energy
IV. A catalytic converter contains a homogeneous catalyst

A. I and II

B. II and III

C. I, II, and III

D. II, III, and IV

E. I, III, and IV

The answer is B.
Catalysts provide an alternate mechanism in both directions pushing the reaction to proceed faster, but do not alter equilibrium (I is false, II is true). The kinetic energy of molecules increases with temperature, hence the energy of their collisions increases as well (III is true). Catalytic converters contain a heterogeneous catalyst (IV is false).

CHEMISTRY

54. Consider the reaction between iron and hydrogen chloride gas:

$$Fe(s) + 2HCl(g) \rightarrow FeCl_2(s) + H_2(g)$$

7 moles of iron and 10 moles of HCl react until the limiting reagent is consumed. Which statements are true?

I. HCl is the excess reagent
II. HCl is the limiting reagent
III. 7 moles of H_2 are produced
IV. 2 moles of the excess reagent remain

A. I and III

B. I and IV

C. II and III

D. II and IV

E. I only

The answer is D.

The limiting reagent is estimated from dividing number of moles of each reactant by its stoichiometric coefficient. The lowest result is the limiting reagent and the highest is the excess reagent:

$$7 \text{ mol Fe} \times \frac{1 \text{ mol reaction}}{1 \text{ mol Fe}} = 7 \text{ mol}$$

$$10 \text{ mol HCl} \times \frac{1 \text{ mol reaction}}{2 \text{ mol HCl}} = 5 \text{ mol}$$

Therefore, HCl is the limiting reagent (II is true) and Fe is the excess reagent (I is false). 5 moles Fe (limiting reagent) reacts with 5 moles HCl, leaving 2 moles of the excess reagent (IV is true).

CHEMISTRY

55. The reaction: $(CH_3)_3CBr(aq) + OH^-(aq) \rightarrow (CH_3)_3COH(aq) + Br^-(aq)$
occurs in three elementary steps:

$(CH_3)_3CBr \rightarrow (CH_3)_3C^+ + Br^-$ is slow
$(CH_3)_3C^+ + H_2O \rightarrow (CH_3)_3COH_2^+$ is fast
$(CH_3)_3COH_2^+ + OH^- \rightarrow (CH_3)_3COH + H_2O$ is fast

What is the rate law for this reaction?

A. Rate = k $[(CH_3)_3CBr]$

B. Rate = k $[OH^-]$

C. Rate = k $[(CH_3)_3CBr][OH^-]$

D. Rate = k $[(CH_3)_3CBr]^2$

E. Rate = $\dfrac{k\,[(CH_3)_3CBr]}{[(CH_3)_3COH]}$

The answer is A.

The first step will be rate-limiting that will determine the rate for the entire reaction because it is slower than the other steps. As can be seen from stoichiometry of the reaction, it is a unimolecular process where the rate is given by answer A. Choice C would be correct if the reaction as a whole is a one elementary step instead of three, but the stoichiometry of a reaction composed of multiple elementary steps cannot be used to predict a rate law.

CHEMISTRY

56. Which statement about thermochemistry is true?

A. Particles in a system move less freely at high entropy

B. Water at 100 °C has the same internal energy as water vapor at 100 °C

C. A decrease in the order of a system corresponds to an increase in entropy

D. The Heat of Fusion is the energy needed to transform a liquid into a solid

E. At its sublimation temperature, dry ice has higher entropy than CO_2 gas

The answer is C.
At high entropy, particles (molecules or atoms) have a large freedom of motion (choice A is false). Water and water vapor at 100 °C contain the same kinetic energy, but water vapor has additional internal energy that prevent intermolecular attractions between molecules, which makes choice B is false. Entropy is interpreted as degree of disorder or randomness in a system, and the more entropy decreases, the more the system tends to be stable and reaches total order at 0 °K, where S=0 (C is correct). Heat of Fusion corresponds to change in enthalpy that results from heating a substance to change its state from a solid to a liquid (choice D is false). Sublimation is defined as transition of substance from solid to gas phase. As a result, freedom of motion for particles in solids is lower than in gases and solid CO_2 (dry ice) has a lower entropy than gaseous CO_2 (choice E is false).

57. Which statement about reactions is true?

A. All spontaneous reactions are exothermic and cause an increase in entropy.

B. An endothermic reaction that increases the order of the system cannot be spontaneous.

C. A reaction can be non-spontaneous in both directions (forward and backward).

D. Melting snow is an exothermic process.

E. Thermodynamic functions are dependent on the reaction pathway.

The answer is B.
All reactions that are both exothermic ($\Delta H <0$), and cause an increase in entropy ($\Delta S>0$) will be spontaneous, but the reverse is not true (choice A is false). Some spontaneous reactions are exothermic but decrease entropy and some are endothermic and increase entropy (choice B is correct). The reverse reaction of a non-spontaneous reaction will be spontaneous (choice C is false). Snow requires heat to melt, thus it is an endothermic reaction (choice D is incorrect). Kinetic constants depend on the reaction pathways that could be accelerated or decelerated by use of mediators and catalysts (choice E is false).

CHEMISTRY

58. Given:

$E° = –2.37$ V for
$Mg^{2+} (aq) + 2e^- \rightarrow Mg (s)$

and

$E° = 0.80$ V for
$Ag^+ (aq) + e^- \rightarrow Ag (s)$

what is the standard potential of a voltaic cell composed of a piece of magnesium dipped in a 1 M Ag^+ solution and a piece of silver dipped in 1 M Mg^{2+} solution?

A. 0.77 V

B. 1.57 V

C. 3.17 V

D. 3.97 V

E. 0.38 V

The answer is C.
$Ag^+ (aq) + e^- \rightarrow Ag (s)$ has a higher value of standard reduction potential E° (energy) than Mg^{2+} $(aq) + 2e^- \rightarrow Mg (s)$. Hence, Ag^+ should collect electrons because it has highest energy and Mg should give its electrons because it has a lower energy (electrons are easy to remove from its orbitals). Therefore, in the proposed cell, reduction occurs at the Ag that will be the cathode and oxidation occurs at the Mg electrode that will be the anode. The following equation could be used to estimate the cell potential:

$E°_{cell} = E°(\text{cathode}) - E°(\text{anode})$, $E°_{cell} = 0.80V – (-2.37V) = 3.17V$ (Answer C)

Choice D results from the incorrect assumption that electrode potentials depend on the stoichiometry of the oxidation and reduction reactions. The balanced overall reaction of the cell is written as follows:

$Mg(s) \rightarrow Mg^{2+}(aq) + 2e^-$ $\qquad E°_{ox} = 2.37V$
$2Ag^+(aq) + 2e^- \rightarrow 2Ag(s)$ $\qquad E°_{red} = 0.80V$ (not 1.60V)
$Mg(s) + 2Ag^+(aq) \rightarrow 2Ag(s) + Mg^{2+}(aq)$ $\qquad E°_{cell} = 3.17V$ (not 3.97V)

CHEMISTRY

59. What could cause this change in the energy diagram of a reaction (the energy scale is exactly the same for both figures)?

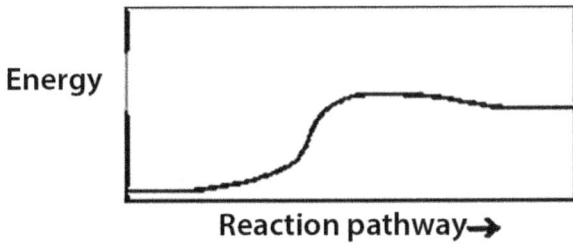

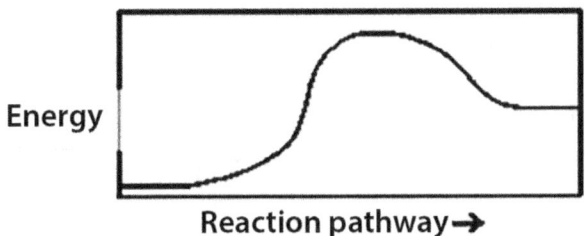

A. Adding a catalyst to an endothermic reaction.

B. Removing a catalyst from an endothermic reaction.

C. Adding a catalyst to an exothermic reaction.

D. Removing a catalyst from an exothermic reaction.

E. Adding heat to an endothermic reaction.

The answer is B.
The products at the end of the reaction pathway have higher energy than the reactants; hence the reaction is endothermic (narrowing down the answer to A or B). The maximum height on the energy diagram corresponds to the activation energy that is caused by removal of a heterogeneous catalyst (choice B is correct). In other words, in presence of a catalyst, the reaction will require less energy to be achieved (A is incorrect).

CHEMISTRY

60. In the following phase diagram, _____ occurs as Pressure is decreased from A to B at constant Temperature and _____ occurs as Temperature is increased from C to D at constant Pressure.

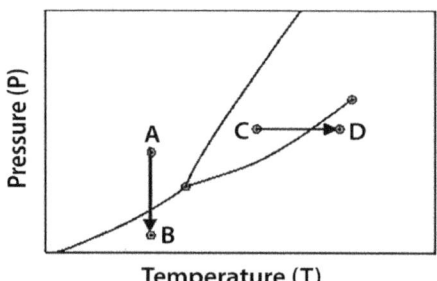

   A. deposition, melting

   B. sublimation, melting

   C. deposition, vaporization

   D. sublimation, vaporization

   E. melting, vaporization

The answer is D.
Point A is located in the solid phase, point C in liquid phase, and points B and D are in the gas phase. The transition from solid to gas is sublimation and from liquid to gas is vaporization.

61. Heat is added to a pure solid at its melting point until it all becomes liquid at its freezing point. Which of the following occur(s)?

   A. Intermolecular attractions are weakened

   B. The kinetic energy of the molecules does not change

   C. The freedom of the molecules in movement increases

   D. The temperature of the system remains constant

   E. All of the above

The answer is E.
Intermolecular attractions are weakened during melting of a solid. This allows molecules to gain more freedom in movement because they are in liquid phase. However, because the temperature remained the same, there will be no change in the kinetic energy of the molecules. Thus, all answers are correct.

CHEMISTRY

62. This compound:

contains an:

A. alkene, carboxylic acid, ester, and ketone

B. aldehyde, alkyne, ester, and ketone

C. aldehyde, alkene, carboxylic acid, and ester

D. acid anhydride, aldehyde, alkene, and amine

E. aldehyde, amine, ketone, and alcohol

The answer is C.
The organic functions are encircled below:

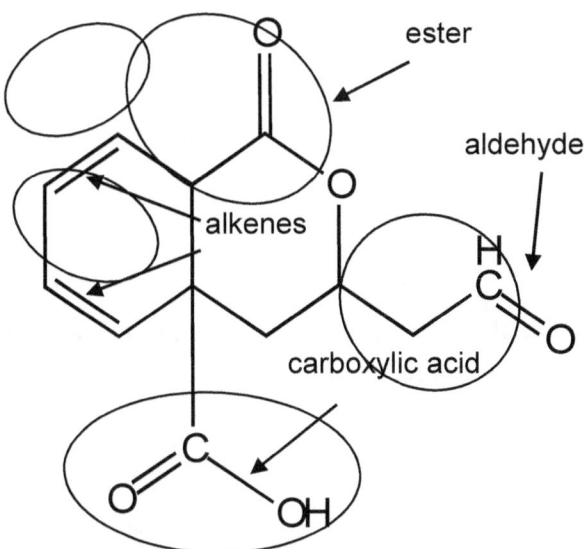

Choice A is wrong because there are no ketones in the molecule. A ketone has a carbonyl group (-C=O) linked to two hydrocarbons to yield (R_1-CO-R_2). Choice B is wrong as well because there are no ketones and no alkynes (-C≡C-) in the molecule. Choice D is wrong because there are no acid anhydrides (-C(CO)-O-C(CO)-) and no amines (-$NR_1R_2R_3$) because amines require at least one N-C bond and no nitrogen atoms are available in the proposed molecule.

CHEMISTRY

63. Which of the following pairs are isomers?

I. [Structure showing two hydrazine derivatives with CH$_3$ and H substituents on N—N]

II. pentanal, 2-pentanone

III. [Two cyclopentane structures with Br substituents]

IV. [Two tetrahedral carbon structures with H$_3$C, OH, H, F and H, OH, F, CH$_3$ substituents]

A. I and IV

B. II and III

C. I, II, and III

D. I, II, III, and IV

E. None of the above

The answer is B.

In Pair I, the N—N bond may freely rotate in the molecule because it is single and not a double bond. Hence, the pair is actually one molecule represented twice.

In Pair II, pentanal is [structure of pentanal] and 2-pentanone is: [structure of 2-pentanone]

246

CHEMISTRY

Both molecules have an overall formula of $C_5H_{10}O$, and they are isomers because they have the same formula but different arrangement of atoms to yield an aldehyde in pentanal and ketone in 2-pentanone.

In Pair III, both molecules are 1,3-dibromocyclopentane ($C_5H_8Br_2$). The bromines are in trans configuration in the first molecule and cis configuration in the second molecule. Since these molecules are rings and cannot freely rotate as single bonds (I), thus they yield different arrangements and are isomers.

Pair IV (1-fluoroethanol) has a chiral center and stereoisomers are possible. However, as in Pair I same molecule is represented twice and rotating the C-O bond results in superimposable structures, hence they are not isomers. Consequently, only Pairs II and III are isomers (B is correct).

64. **Which of the following is not a colligative property?**

    A. Viscosity lowering

    B. Freezing point lowering

    C. Boiling point elevation

    D. Vapor pressure lowering

    E. All of the above

The answer is A.
Colligative property depends on ratio of the number of solute particles to the number of solvent molecules in a solution. Vapor pressure lowering, boiling point elevation, and freezing point lowering may all be visualized as solute particles in between two phases where dilution or concentration of solute particles may occur. This is not the case for viscosity.

CHEMISTRY

65. A sample of 50.0 mL KOH is titrated with 0.100 M HClO$_4$. The initial buret reading is 1.6 mL and the reading at the endpoint is 22.4 mL. What is [KOH]?

   A. 0.0416 M

   B. 0.0481 M

   C. 0.0832 M

   D. 0.0962 mM

   E. 0.0962 M

The answer is A.
HClO$_4$ and KOH are both strong electrolytes (acid and base, respectively), hence they fully dissociate in water as follows:
HClO$_4$ → H$^+$ + ClO$_4^-$ and KOH → K$^+$ + OH$^-$).

The problem could simply be solved using the dilution formula ($C_{known} V_{known} = C_{unknown} V_{unknown}$):

$$C_{unknown} = \frac{C_{known}(V_{final} - V_{initial})}{V_{unknown}} = \frac{0.100M \, (22.4mL - 1.6mL)}{50.0mL} = 0.0416M$$

The problem may also be solved by finding number of moles of the known substance:

$$0.100 \frac{mol}{L} \times \frac{1L}{1000mL} \times (22.4mL - 1.6mL) = 0.00208 \text{ mol } HClO_4$$

This will neutralize 0.00208 mol KOH, and $\frac{0.00208 \text{ mol}}{0.0500 \text{ L}} = 0.0416M$

CHEMISTRY

66. **When KNO₃ dissolves in water, the water grows slightly colder. An increase in temperature will _____ the solubility of KNO₃.**

   A. increase

   B. decrease

   C. double

   D. have no effect on

   E. have an unknown effect with the information given on the solubility.

The answer is A.
The decline in water temperature indicates that the reaction of solubility of the salt KNO₃ in water is endothermic (consumes heat). An increase in temperature will supply heat to the system that, in turn, will favor an increase in solubility according to Le Chatelier's principle.

CHEMISTRY

67. Classify these biochemicals:

I.

II.

III.

IV.

A. I - nucleotide, II - sugar, III - peptide, IV – fat

B. I – DNA, II sugar, III- peptide, IV lipid

C. I - disaccharide, II - sugar, III - fatty acid, IV-polypeptide

D. I - disaccharide, II - amino acid, III - fatty acid, IV - polysaccharide

E. I - nucleotide, II - sugar, III - triglyceride, IV – DNA

The answer is A.
Compound I is a phosphate (PO_4) linked to a sugar and an amine, this is named nucleotide. Compound II has the formula $C_nH_{2n}O_n$, indicative of a sugar. Compound III contains three amino acids linked with peptide bonds, this is called a tripeptide. Finally, compound IV is a triglyceride, characterized as fat molecule.

CHEMISTRY

68. Which of the following can be determined from the Periodic Table?

    I. The number of protons
    II. The number of neutrons
    III. The number of isotopes of that atom
    IV. IV The number of valence electrons

    A. I only

    B. I and II

    C. I, II, and III

    D. I, II, III, IV

    E. I, II and IV only

The answer is E.
The number of protons, neutrons and electrons can be determined from the Periodic Table. Isotope varies only in number of extra neutrons within an atom.

69. Which of the following quantum numbers are needed to define the position of the electrons in an element:

    A. principal, angular momentum, magnetic, and spin

    B. principal, circular, magnetic and electromagnetic

    C. angular, magnetic, electronic, spin

    D. primary, angular momentum, magnetic, and spin

    E. principal only

The answer is A.
All four quantum numbers: principal, angular, magnetic and spin, are required to fully define the position of an electron in an element.

CHEMISTRY

70. Which of the following are true:

A. Anions are larger than their corresponding atom.

B. Second Ionization Energy is greater than the first Ionization Energy.

C. As you move down a group on the Periodic Table the atomic radius increases.

D. Atoms with completed shells are more stable.

E. All of the above are true

The answer is E.
Anions contain extra electrons; hence they become larger in size if compared to neutral atom. Removal of a second electron from an atom requires more energy than the one utilized to remove the first electron because electrons of lower orbitals are more stable than electrons at the periphery or in higher orbitals. As you move down a group on the Periodic Table, electrons are added and atomic radius increases. Finally, Atoms with completed shells are more stable because they have fewer tendencies to participate in chemical reactions. Consequently, all of the statements are true.

71. The solubility of $CoCl_2$ is 54 g per 100 g of ethanol. Three flasks each contain 100 g of ethanol. Flask #1 also contains 40 g $CoCl_2$ in solution. Flask #2 contains 56 g $CoCl_2$ in solution. Flask #3 contains 5 g of solid $CoCl_2$ in equilibrium with 54 g $CoCl_2$ in solution. Which of the following describes the solutions present in the liquid phase of the flasks?

A. #1 - saturated, #2 - supersaturated, #3 - unsaturated.

B. #1 - unsaturated, #2 - miscible, #3 - saturated.

C. #1 - unsaturated, #2 - supersaturated, #3 - saturated.

D. #1 - unsaturated, #2 - not at equilibrium, #3 - miscible.

E. #1 - unsaturated, #2 -- saturated, #3 - miscible.

The answer is C.
Flask #1 contains less solute than the solubility limit which makes it unsaturated. Flask #2 contains more solute than the solubility limit which makes it supersaturated. Flask #3 contains the solubility limit of a saturated solution at equilibrium with the solid salt. The term "miscible" applies only to liquids that mix together in all proportions.

CHEMISTRY

72. An experiment requires 100 mL of a 0.500 M solution of MgBr$_2$. How many grams of MgBr$_2$ will be present in this solution?

A. 9.21 g

B. 18.4 g

C. 11.7 g

D. 12.4 g

E. 15.6 g

The answer is A.
Use of molarity (c = m/MV $\Rightarrow$ m = VcM).

$$0.100\text{L solution} \times \frac{0.500 \text{ mol MgBr}_2}{\text{L}} \times \frac{(24.305 + 2 \times 79.904)\text{g MgBr}_2}{\text{mol MgBr}_2} = 9.21\text{g MgBr}_2$$

73. Which of the following is most likely to dissolve in water?

A. H$_2$

B. CCl$_4$

C. SF$_6$

D. CH$_3$OH

E. CH$_4$

The answer is D.
The best solutes for a solvent are those with intermolecular bonds of similar strength to the solvent. H$_2$O molecules are connected together to form liquid network (water) thanks to hydrogen bonds. H$_2$, SF$_6$, and CH$_4$ are gases with poor solubility in water due the intermolecular attractions in these gases that are mainly governed by weak London dispersion forces. CCl$_4$ is a liquid that dissolves poorly in water because it can not form hydrogen bonding with water molecules, but its solubility increases in organic solvents alcohols, chloroform, ethers …etc. Finally, CH$_3$OH (methanol) is highly miscible with water because it can form hydrogen bonding through –OH groups with water molecules H$_2$O.

CHEMISTRY

74. 10 kJ of heat are added to one kilogram of iron at 10 °C. What would be the final temperature? The specific heat of iron is 0.45 J/(g.°C).

  A. 22 °C

  B. 27 °C

  C. 32 °C

  D. 37 °C

  E. 14.5 °C

The answer is C.
The expression of heat as a function of change in temperature is written as follows:
$$q = n \times C \times \Delta T$$
where n is the mass, C is the specific heat of iron, and q is the heat. This expression might be rearranged to determine change in temperature:
$$\Delta T = \frac{q}{n \times C}$$

$$\Delta T = \frac{10000 J}{1000g \times 0.45 \frac{J}{g \cdot °C}} = 22 \text{ °C}$$

Note that ΔT is not the final temperature (choice A is incorrect), but difference between the initial and final temperature. $\Delta T = T_{final} - T_{initial} = 22$ °C

Hence, the final temperature should be: $T_{final} = \Delta T + T_{initial} = 22$ °C + 10 °C = 32 °C (Choice C)

CHEMISTRY

75. A Student wishes to prepare 4.0 liters of a 0.500 M KIO₃ (molar mass 214 g). The proper procedure is to weigh out:

A. 42.8 g of KIO_3, and add 4 kg of H_2O.

B. 42.8 g of KIO_3 and add H_2O until the final homogenous solution reach a volume of 4.0 L.

C. 21.4 g of KIO_3, added to 4 L of water.

D. 42.8 g of KIO_3 added to 4 L of water.

E. 214 g of KIO_3 added to 4.0 L of H_2O.

The answer is B.
Always weight the salt first, and then add liquid until the desired volume. The reverse process could cause problems such as errors in estimated final volumes (concentration).

Use of molarity (c = m/MV ⇒ m = cMV):
0.5 (moles/liter) x 214 (g/mole) x (4.0 L) = 428 g

As a result, 428 g of KIO_3 should be dissolved in water, then water should be added to the solution until the final volume reaches 4.0 L. This means that the added volume is not necessarily 4 L of water, but could be less since the salt takes already some volume in the solution.

XAMonline
The CLEP Specialist

Individual Sample Tests in ebook format with full explanations

eBooks

All 33 CLEP sample tests are available as ebook downloads from retail websites such as **Amazon.com** and **Barnesandnoble.com**

American Government	9781607875130
American Literature	9781607875079
Analyzing and Interpreting Literature	9781607875086
Biology	9781607875222
Calculus	9781607875376
Chemistry	9781607875239
College Algebra	9781607875215
College Composition	9781607875109
College Composition Modular	9781607875437
College Mathematics	9781607875246
English Literature	9781607875093
Financial Accounting	9781607875383
French	9781607875123
German	9781607875369
History of the United States I	9781607875178
History of the United States II	9781607875185
Human Growth and Development	9781607875444
Humanities	9781607875147
Information Systems	9781607875390
Introduction to Educational Psychology	9781607875451
Introductory Business Law	9781607875420
Introductory Psychology	9781607875154
Introductory Sociology	9781607875352
Natural Sciences	9781607875253
Precalculus	9781607875345
Principles of Macroeconomics	9781607875406
Principles of Microeconomics	9781607875468
Principles of Marketing	9781607875475
Principles of Management	9781607875468
Social Sciences and History	9781607875161
Spanish	9781607875116
Western Civilization I	9781607875192
Western Civilization II	9781607875208

 or amazon or BARNES&NOBLE BOOKSELLERS

XAMonline.com

XAMonline
CLEP
Full Study Guides

CLEP College Algebra
ISBN: 9781607875598
Price: $34.95

CLEP Biology
ISBN: 9781607875314
Price: $34.95

CLEP Analyzing and
Interpreting Literature
ISBN: 9781607875260
Price: $34.95

CLEP College Composition
and Modular
ISBN: 9781607875277
Price: $19.99

CLEP College Mathematics
ISBN: 9781607875321
Price: $34.95

CLEP Psychology
ISBN: 9781607875291
Price: $34.95

CLEP Spanish
ISBN: 9781607875284
Price: $34.95

 TO ORDER ▸ **X** XAMonline.com or **amazon** or **BARNES & NOBLE** BOOKSELLERS

XAMonline
CLEP Subject Series
Collection by Topic
Sample Test Approach

CLEP Literature
ISBN: 9781607875833
Price: $34.95

CLEP Foreign Language
ISBN: 9781607875772
Price: $34.95

CLEP History
ISBN: 9781607875789
Price: $34.95

CLEP Sociology
ISBN: 9781607875796
Price: $34.95

CLEP Science
ISBN: 9781607875802
Price: $34.95

CLEP Mathematics
ISBN: 9781607875819
Price: $34.95

CLEP Business
ISBN: 9781607875826
Price: $34.95

 TO ORDER or amazon or BARNES & NOBLE BOOKSELLERS

XAMonline.com

XAMonline
CLEP Favorites
Collection by Topic
Sample Test Approach

CLEP Five Favorites
ISBN: 9781607875765
Price: $24.95

CLEP Military Favorites
ISBN: 9781607875512
Price: $24.95

www.ingramcontent.com/pod-product-compliance
Lightning Source LLC
Chambersburg PA
CBHW080728230426
43665CB00020B/2665